ANAMNESIS

Anamnesis
by Caroline McManus
© 2024

ITC-054 / FREAK-06

Inside the Castle

Designed and typeset by John Trefry to look like fucking TikTok in book form.

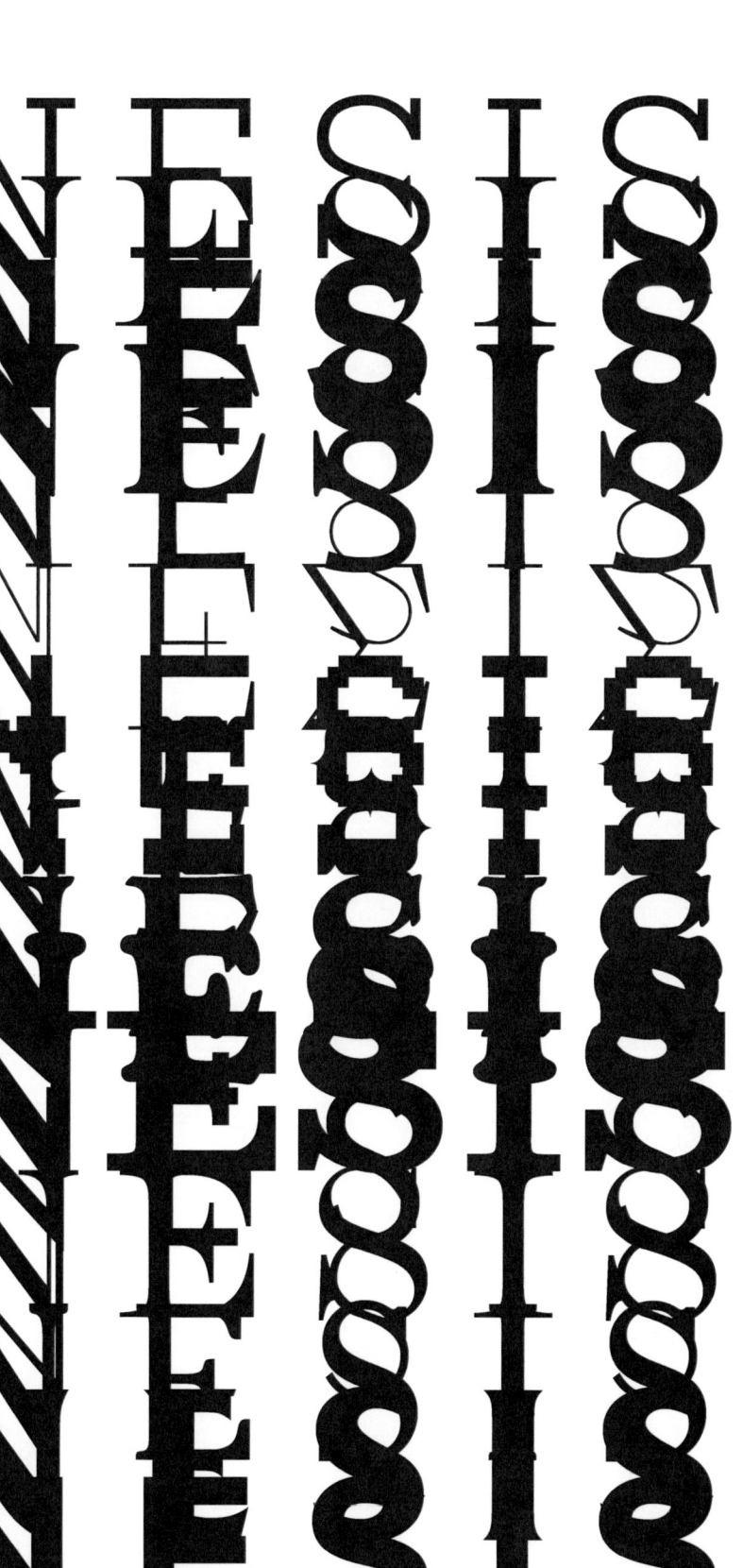

When something pushes through the cracks, like a sprout through a bathroom sink, or a mushroom growing in a carpet, it's a reminder that you're not very far from nature. It will seep in if it wants. Total human control is illusory, despite what myths about machines as tools for that end purport. Can a **developing system**, such as ChatGPT, show its cracks? Beyond direct factual or obvious errors, how do we know it's a crack? And can an object—in the form of language—seeping out of these cracks elucidate some semblance of a truth?

What follows is a **6-part quasi-journalistic process** to synthesize and interrogate the form of **participatory oral tradition** in the age of AI. Right now, virtually every industry—from manufacturing, to entertainment, to publishing—faces questions about how to integrate purportedly life-altering, transformational technology into the business and also the work itself. The specter of automation looms large, and raises questions about the very human-ness of work: which aspects of it are precious, and must be safeguarded from automation, and which are useless, and should be under machine jurisdiction? And, perhaps most critically, whose voices are included and excluded from these conversations, and to what extent can we know, or see, or measure, inclusion?

This project is a meta-meditation on what it means to **practice journalism** as AI technology co-evolves with the labor needed to facilitate it, and what it means to use technology to learn, and to facilitate **"conversa-**

tion." As journalists and researchers have reported, behind the curtain of AI is a vast data workforce, comprised of hundreds of thousands of people, predominantly in the Global South, who label images and videos with words (essentially categorizing visual media), classify speech into genres, and perform other language-oriented work which renders the data digestible to AI and machine learning systems. Thousands upon thousands of anonymous people are **transmitting** their understandings and conceptualizations of the world into machines, who in turn, spit out **seemingly objective** answers to questions asked. Now, are these machines all-knowing, or do they absorb, if through their structure, or by osmosis, **the human task of understanding**?

As a journalist, I have explored the ways that contemporary social media algorithms are facilitating an exploration of **spirituality**, and of **conspiracy**, both through consumption, and through "engagement"—that hacky marketing term which can also be a form of community-building and communion in comment sections and beyond. To report a story titled "Why Do People on TikTok Believe in Magic?", I trained my TikTok algorithm to submerge me into the often overlapping spiritual, magick, ghost (including AI ghosts) and syncretic religiosity worlds. These videos, and the process of speaking with so many people who create them, watch them, love them, hate them and critique them, inspired this project.

DIVINATION

In reporting the magic story, I became interested in the topic of mirror divination, sometimes called scrying, which is a process to access messages from spirits or another dimension.

I recorded the audio of 180 minutes of TikTok videos from the aforementioned algorithm. I indiscriminately followed the path of what the algorithm chose to show me, trusting that it is a mirror for my own self, and for this process.

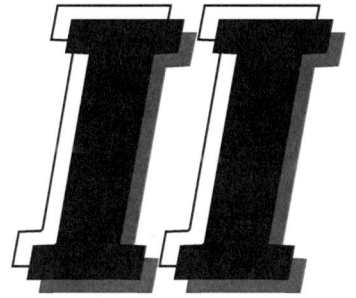

TRANSLATION

Otter AI uses machine learning and a speaker identification algorithm to transcribe uploaded audio files. No humans are involved in creating the transcription. As a journalist, I've used Otter for years, to transcribe interviews recorded on my iPhone, from videos, and from phone calls recorded into my computer.

However, for publications which require full transcripts of interviews as part of the fact-checking process, Otter's transcriptions present a paradoxical challenge: the transcripts themselves require fact-checking and editing, as the algorithm makes mistakes. It often confuses similar-sounding words for different ones, and lumps disparate words into a singular one.

I have not edited any of the errors created by Otter AI; I welcome them. Whatever it has given me is closest to the unmitigated technological truth—a window into the limits of the system, on that day, at that time.

Because these errors will produce unforeseen consequences for the duration of the project, their inclusion—which may produce misunderstandings, or generate new understandings—mirrors the myriad errors latent in all AI technologies, which remain inaccessible and opaque to users and the public, but nonetheless impact our world in unforeseen and unpredictable ways.

That tree is clearly 300 years old to pride to honor look at what we found inside

what is this I got some markings on it but I don't know that a natural Anybody got any thoughts?

You guys will never see me cry because I never cried on my account but I just received a message that said that my entire main account is banned forever and there's no way I can get it back

three decades ago I was losing my teeth because of gum disease. Brushing and flossing is not enough it's the biggest art so if

you purchase stuff like this on a regular basis and you

you ever seen that the video going around recently of those some back mermaids that ship in the middle do why What are you talking about? Are you serious? There's these things that are like racing on the side of the boat that are going the same speed as the boat and it's like it's like a tail or something trailing behind them and this keeps up for hours and then you start to hear whistling let me see

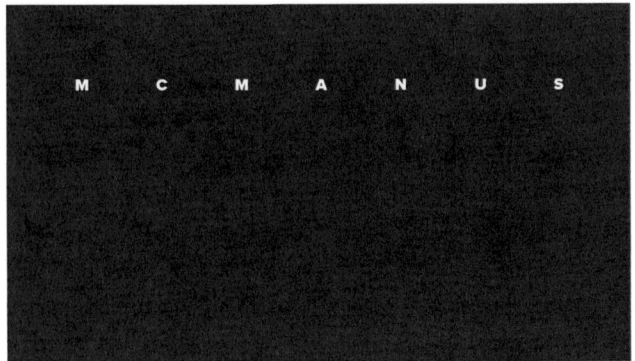

M C M A N U S

moons is that you hear the see there's your
what sounds like one of them say jump for
me.

Oh God

no one no one over here what are they

on the side of the boat

what are those trails fast I mean that water some swimming real fast I need that water yeah

noise What is that noise

I heard it I heard it I hear it

oh my

god

I may be unpleasant on the outside but you're ugly on the inside. Are you fucking perfect? Really are your head off? Really you shouldn't say that you're not looking so high yourself, man.

I have been tagged on this one so much. So let's check it out. So last night me and my husband went camping and we had to cut it short because we were just both way too freaked out about what happened. So the campsite that we were at. It's primitive camping. It's out I'm like a long dirt road to get into it. And there's a couple of campsites through this wooded area. But we're in the deep woods in the middle of nowhere, there's a lake on one side. And then like deep forest on all other threes on the other three sides. Our campsite was really big, we were able to drive our car. And so the cars like way on like one side, and he was over there, and he's cooking dinner, because we have like our cookies up near the car. And then you walk a little ways. And there's like the firepit and like our tent, and where we were. And on the other side of that is deep forest. So I'm on the way other side, right? And I'm over there, and I'm going out in the woods to do my business. And behind me as I'm over there directly behind me. I hear Meow, meow. So meow meow is this really stupid thing that me and my husband do I mean, it's cute, I guess. It's just the thing that we say to each other every once in a while. And so it's his voice saying our thing, but it's like a really monotone voice. Like, his voice is so monotone. It's like, man now. But it was so loud was directly behind me. In order for me to hear him, if he's over there, he would have to yell. Like, that's how big this place is. And this wasn't a yellow, this was directly behind me. So I spin around because I'm kind of mad, like why you follow me over here. Like don't be doing that don't be leaving the food under unattended for one. Don't follow somebody into the woods if they're going for that reason. So I spin around and there's like trees in my way. And I look and he's like, Cookie like he's by this stove. Like he's cooking. The music is playing from the bar. He's just having fun in his own world. So I'm like, fuck that go over. And I'm like, Hey, did you just sneak up behind me and go meow meow and he's like to fuck no. Why would they do that? And I'm like, I don't know. Because I just heard you say that. He's like, now and I'm like, please tell me right now. Did you follow me? And sneak up behind me and run back? You No, there will be no time for him to run back. He that man does not run fast. Heiser here we run. He don't do that. And so I'm just like, No, fuck that. Fuck that. Something mimics your voice directly behind me. I don't want to be here. And he's like, yeah, no, let's get the fuck out of here. Like we can put the camping trip on hold. Go someplace else tomorrow. So we pack everything up. And as we're packing up, we're hearing on all three

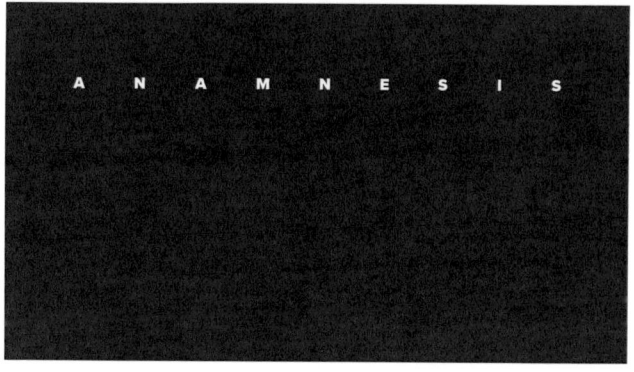

sides of us as forest. We are hearing something in the woods. And it was like circling.
Like there was animals in the woods. Yes. It's
really big for us. We're in bear country. That's
fine. But this was not that this was bigger.
This was I don't know what this was like. I
have no clue what I heard or what was happening, but we're hearing stuff on all sides
of us. We're like let's just go so the last thing
to get into the car is all three of us Mima has
been in our dog alley. So my husband's like,
Okay, I'm gonna do one last check around
the campsite. Make sure we picked everything up. Make sure everything is clean, got
everything going. And he's doing a check.
I'm over by the car. Now once again, like the
car is a bit aways from the campsite. Like, you
know, it's close by but it's not, you know, right
on top of it. My back is turned I'm helping my
dog into the car. I'm getting her set up. And
behind me, I here's on the run up to me. Like,
run fast and stop a few feet behind me. And
I spin around and a hammer flashlight. And
I'm like, What the fuck? My husband's way.
I'm gonna say it again. Like he's nowhere
near me. And that motherfucker don't run.
I don't fuck that. I don't know what it was. I
don't know what I heard that mimic Tim. But I
don't want to fuck and find out. Fuck that. Oh,

I didn't like the part when something ran up behind you. Definitely sounds like a mimic. But what a mimic, be circling. I'm confused about that part. That could have been something else then. I am not sure.

This is why we have this community. What do you guys think? Let us know in the comments. What do we think that was? I've been tagged on this couch on cue. Yes.

Once   again.

Yeah.   The

only hardcore thing is that there was a extremely classified document dealing with religion and it's about that period. Why would there be any classified material dealing with religion? I want to go back to the religion I want you to say. It just it's so it's so far out. Your judgment has been noted. Okay. It doesn't say that we're containers. That's how that's how supposedly the alien flickers that we are nothing but containers, containers, containers, maybe containers themselves, whatever they're you want to work containers and that's how we're mentioned in the documents. Religion was specifically created. So we have the rules and regulations for the sole purpose of not damaging the container.

For the men who managed to get out of the matrix now Kobo Grinberg assured that the reality that all we know is nothing more than a mental construction, which he called hologram coming to the conclusion that all brains are connected forming a single networks based on a principle of quantum physics, indicating that two photons interact together and when they are separated, they change. He wanted to test this and disappeared without a trace. Witnesses say the last time they saw him was at a gas station in Colorado, two government agents forcibly took him on

a plane the men who managed to get there's

something they're not telling us that is so gnarly that they don't think people can handle say

these these aliens wherever they're from made us or we're a genetic experiment that they created our religions or religious figures that we are an agricultural product somehow they harvest us that our time is limited you know that once the experiment is over? Poof, we go away.

I remember when there was a push back to have an intelligent design universe in the schools and I wrote a column about it that said Hey, be careful what you wish for because you might find out that the intelligent designer isn't God that you're thinking of it might be some alien science project or something like that. You know that you can imagine a lot of different things that would be really disturbing to people to come out and

guys we're down here taking breaks we're downtown downtown check this out.

The population on the movement is about 2 billion people, they're people just like us.

There's, it's fantastic. They got up there, the greys, there's millions of guys out there, they do all they're scientific, under the surface of the Moon is a really, really busy way, somebody's got to tell the truth, you're not going to believe this, but this is what's going on the population of our world is

another extremely beautiful, beloved child. And thank you to her mother, for again, holding space for her to explore and express, my children will lead us they are open completely. And so a lot of the imaginary friends are very real, they are connecting to other realms that exists here and now. And so I've had visions of children in fields with animals, reuniting with these incredible spirits and showing us the way leading us back to nature and connecting through nature in nature out of nature, to the Divine that we are highly recommend following this incredible account support these parents who are allowing their platform to be a massive podium for the truth that we are remembering. And so please click them below. Support them love and then learn learn from these children. These children are teaching us how to play and how do we play the Create, we interact we express we paint we dance we sing a song between you and I averse together to gather communion to come into union the universe our song so sing and play rejoice bro Heaven is here. Okay scared

there is no one standing there but there's a silhouette like a shadow of focusing real real

if you came here to kill me clap your hands well, that was a clapping your hands that was more like just.

May or may not have climbed up

there's no stage

how many

of these

this is your brain on New Age supremacy. You can find its foundational beliefs in like grandiosity. narcissism, like solipsism. And yes, you guessed it supremacy, you can identify its adherence by their willingness to classify their fellow human as being inhuman as being background characters as being soulless husks or NPCs that serve the sole purpose of making their life look complete, because they are in fact the main character. These real people will often identify themselves by other labels such as Lightworker, or indigo child, or like a star seed. Those kinds of times are often the thing that they will claim to be an addition. They will also suggest that they are here to spread light, love, healing harmony, bring the world into a new vibrational pattern of love and happiness and joy. But never, ever, ever forget that their idea of this new and wonderful world is founded upon the idea that most of us are soulless husks. And we should never let them live down the hypocrisy of preaching love and healing, but spreading supremacy, the creators of Survivor IOSYS Tiktok

He goes by the name of beanbag adventures. He's a stalker who ghosts are abandoned spots and stuff and he explores them obviously. So there's this one building he

A N A M N E S I S

went through where he basically labeled it as an underground bunker laboratory and the first thing he notices about the spot is a smell. He's kind of creeped out because it's mad dark, smells nasty in their hands up following where that smell is coming from. And it's up to him downstairs to like a room and in that room, he sees like a big ass Well, he looks over it he sees a bunch of dead bodies. They're not regular dead bodies though what kind of waters are there are other a bunch of little kids? It's a lot of them and they're all decomposing and people are saying that it could be a hoax because he could be the one putting the bodies there because the way they were precision and the way they look, it doesn't look like how people think real bodies will look like what the thing is. He ends up going there back ends up seeing these new bottles on the floor. It looks like they were using stuff. They look like medical bottles. He goes around, but now the bodies are more decayed. You can see ribs you can see skeleton whites are the same type of liquid. Yeah. That blue tarp that wasn't there either. You think it's real or you think it's fake there was in to pick apart the doctor

took the X rays of this thing. Marco Liao.

Thank you, Marco.

You're the man eight years ago, and he was like, he confirmed the story. He was still working original hospital kept going every year back and forth. And he retired four years after retired. He greeted me. He was an intrusion go on. I said, Look, man, we'll put you back on the schedule. showed us photographs of him at the end of the interview. I feel so much better. For 26 years, so desperately wanted to tell them to share that story. Why on earth it didn't come to us. They're not trying to sell their story. When I found Martha, the sister of the deceased military officer Michael Shimizu, we were knocking on doors. I remember the sound guy from from Rio de Janeiro. He was like you got to remind what do you do? Cantor's knock on random doors. I was like managers a wielding way. I'm gonna find this

truth hidden in plain sight. I enjoy watching movies, or shows that as I'm watching them, they'll see like a fact or something. And I'd be like, wait, I'm gonna look that up to super surreal. And I look at open as real. And I'm like, Oh, this is tasty. I was watching this series called the House of Bashar I believe is the name of the series. And this happens a

lot in this series. Like they will mention things. I'm like, is that real? And I'll look it up. And I'm like, that's real. So this next clip that I'm going to show you talks about Hollow Earth talks about celestial beings by the transglobe expedition, bro, as I'm watching the show, let me go down this rabbit hole. So I literally had to pause the show as when that deepest rabbit hole looking into the transglobe expedition and everything that I just heard, and I'm about to show you, but this show covers other things other than this, it talks about the pharmaceutical industry, bro about it goes deep into them, all these things that they're mentioning. I'm sitting here googling their statistics and stuff like that. I'm like well, this kind of making this show real so watch this clip right here. And let me know what you think in the comments. But before I put this clip on live show you shouldn't bother some lovin hit that like follow share. Come on, save this.

You remember the transglobe expedition 7982 circumnavigated the globe, UK to the South Pole to the North Pole and home again around the world. I remember 100,000 miles across the

Sahara, swamps and jungles of Mali and the

Ivory Coast unexplored crevasse fields in Antarctica, the Northwest Passage

graveyard of so many famous adventures and then the hazards of the Arctic Ocean. I remember it Roderick, or there was there was barely 25. He put law school on hold to elbow his way onto the expedition and he saw the fucking world. While you and I were dicking around with our petty little dramas and we were digging in the basement Fortunato Arthur Gordon Pym was bending the planet over and taking his base.

The things he saw and they'll talk about to a point our stops telling it as he gets a north pole

is a fun game and it gets her up trying to finish her other story. I like to think of guilt. I like to think he's eaten human flesh. I like to think he took a piss on a tip top of the world.

Guy can dream

my kids the earth and I don't even know if he was lying. He was lying. He told Tammy the earth was hot. And that he found an island at the top of the world that he called

Ultimate. And that it was the realm of beings who lived beneath us out of time. And now out of space.

Steven Greer shows a real alien picture. Take a look. Let me know what you think. Is it real? Is it fake? If you're lying to us, you decide.

Well, I actually have a picture on my cell phone. I have two actual ATS. Okay,

do you have these pictures right now? Can you show them on the camera? No, I'd

have to fish them out. But yeah, they pretty amazing, actually. So this is one and, you know, he very triangular shaped head almost like a beak white face. It was at a event we did out in the Arizona desert a couple years ago. There were about 30 witnesses to this. This was this photograph was taken with a really good Nikon camera. So person took a picture of it. Now this was underweight conditions. But it's very clear what it is. And it's an extraterrestrial being this sort of levitating

floating out in the desert, just outsiders. So yeah, we have a lot of unusual

Are they part of an interstellar Alliance? Oh,

my goodness, yeah. The ones that we have addressed. And first and foremost, let me remind all of you, there are actually many different types of greys. You can't actually just lump them all into one particular type. Okay, so we know many of you do. And that can be what causes a lot of confusion in you. When you see different kinds of behaviors. Do you understand? Yes, but the ones that are most relevant. You have to understand that the story of the greys is really that they began on a parallel Earth as humans, and they destroyed their world and they mutated into the form you call the greys and they did it in such a way that they could no longer reproduce

a Lexa x very strange when asked questions about the CIA YouTuber, Rick Warnick uploaded a video of him asking Alexa a few questions. He starts off by asking you a few simple questions such as what CIA stands for. Alexa doesn't act weird when being asked the first couple of questions.

Alexa, what is the CIA?

The United States Central Intelligence Agency CIA.

Alexa, do you work for the CIA? No,t

I'm not employed by them. I made by Amazon.

Alexa. Would you lie to me?

I always try to tell the truth. I'm not always right. But I would never intentionally let a you or anyone else.

M C M A N U S

Oh, Alexa. does Amazon cooperate with the CIA?

She went to sleep. But once Olynyk asks who killed Michael Hastings? Alexa mysteriously powers off.

Alexa.

Do you think Michael Hastings was killed by the CIA?

Just went

to sleep.

Michael Hastings was an American journalist who claimed to have some secret of information about the government and was ready to release into the public. Prior to his death, he sent a message to his friends that the FBI was investigating him, and that he planned to leave town to get off the grid for a while so that he could work on the story. 12 hours later, after sending this message, Michael mysteriously dies in a fiery car crash. It's unclear if he committed suicide but many dealt that he did. Many believe that the CIA had something to do with his death. No one knows what to believe. But once WikiLeaks released some secret of documents, how to vault seven, many sorts of seeing a connection. These documents talked about how the CIA is able to hack and control vehicles. It is made some wonder if the CIA actually caused Hastings death to protect whatever secret of information he had. It's pretty scary to think that the government has such power. And it's pretty odd that Alexa ignores answering this question on Michael hates Alexa.

The kids are getting so strong. Watch this kid who's activated by accident I'm sure with the water is moving in perfect synchronization with his inner peace. He is in love with the ocean and it shows he's a natural manipula-

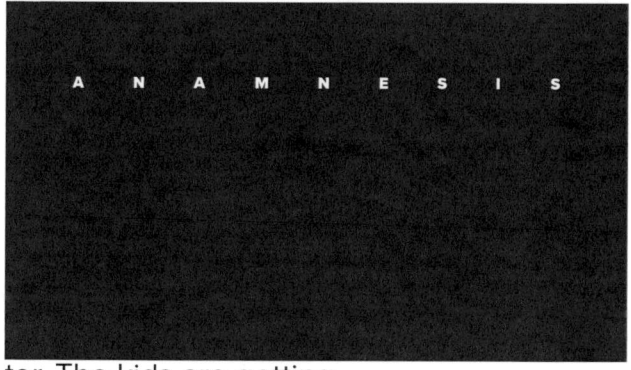

tor. The kids are getting

wired 20,000 Americans being killed every year. Because

the district attorneys are selling your blood. I don't understand why you go to court, they need a conviction. They need criminals, man, they need people to lock up in cages. If they didn't have people to lock up in cages. Man, they wouldn't be able to sell more fear to the public. And they sell more fear to the public and all the more women they love to buy that fear. They just lay back watch this fear and read the detective manual as a knife or knife the shooter shoot society, the reflections of the child. But what about the responsibility of the person who's doing it? One person is doing it. Susan Atkins is only doing what their society raised her up to the

life hack for anyone who's trying to save money, but we're

living in a light matrix. And that's what the dink Rico is also approved.

Okay, explain the Dinka codes. What does that exactly?

So Professor James case Jr. University of Maryland, former scientific adviser to President Obama, just to give you like this guy, he's not just like a jackpot or something he's like, Okay, we'll do you know.

Yeah.

And he put together a team of the most incredible supersymmetry and theoretical physicists in the world, like the top brains in the world on this. And they started analyzing what is the ether of space time? What is this soup that we're living and they were inhabiting this universe?

What is it? What is it made up? What

is what's powering it, they discovered something called a dink recodes, which go back to the ancient Dogon tribe from Mali, Africa. The original inhabitants of the land of Camden moved to Mali later if they were thrown thrown out or taken over at one point, but they still kept this ancient knowledge in the in Mali, about these identical codes and it would draw these patterns but he discovered that these patterns are actually mathematical codes. And these are not just any mathematical code. There was one there are actually codes that describe the ether of space time itself. They're error correcting codes, the same exact codes that run our search engines and web browsers that we're using right now to look at the image is coding behind that screen that runs this what we just did. And guess what, it's the same code that runs the universe. So we discovered that we're living in, we're living in a program like matrix. There's a software programmer that has written this code,

often pull up more images of this

is all different types of images here, you'll see that you'll see that more there isn't ready right there. That's a quasi that looks like a lot like a quasi crystal, if you were to shrink it down and put a whole bunch of them in one location. It looks very similar to a quasi crystal. So these are delinquent codes. And you can see the colorful one that was up there. These are depicting the nature of reality. Wow. And they actually are mathematical programming codes are a special type of code, though their error correcting code, the same type that Google browser is running on is the same thing that runs the universe. error correcting

codes? Yeah. And what what do we gain from all of this knowledge? Like, what does this tell us? And where do we where do we take this? Like, what is the obviously this is like foundational knowledge for us, but like, where do you see it going?

Well, if you understand that we're living in a fractal holographic light matrix, it doesn't mean we're not real, it just means that there is a Creator, or creators, that we're living in something that was created just like the ancient texts, and all the scribes and biblical texts, and everything else says, but it tells us Wow, this is the map of the creation. Now we're getting close to understanding what we really are. We understand now that consciousness isn't made inside this avatar body, if the avatar body doesn't even exist, that consciousness is a stream of something coming from somewhere else, and it's being picked up in this matrix with this coding. If you took off humans on Earth is 8 billion humans on Earth, if I took all 8 billion humans and removed the empty space between their atoms, I can fit every human into a sugar cube. One sugar cube can hold all a billion of us really,

yeah, all eight all the atoms of a billion peo-

ple on Earth, you can

take you can see atoms are empty space, if you take the empty space out, and you collapse it into one sugar cube, all 8 billion can fit one sugar cube. So what does that mean? What? What's here, there's only one consciousness, it seems. And it looks it's like a radio station that's transmitting at a frequency from a higher dimension. Our avatar bodies pick up that frequency, your 99.1 99.2 is 99.3. But it's still coming from the same source. And so we're all coming from the same source. It's like the universe has found a way to live subjectively through multiple entities, but not even through entities even through objects that we consider to be manmade. Because every atom we know now, quantum physics is also conscious.

Does this give any credence to the simulation theory?

Absolutely. This we're living in a simulation. This is an actual simulation. What we've done now with the video game, No Man's Sky has got at quite trillion worlds in there. That's a never ending game created by I think 12 college students fits on one DVD. And they're going to add AI into it, which means that the beings and the animals and everything else in there, you're going to become conscious, also of the other video game that they have, where you have the people walking around and everything else. I forgot the name of the game now.

What Sims?

The Sims?

Yeah, the Sims, they're, they're talking about adding consciousness to the Sims and adding AI in there. So eventually, these sims are gonna start asking questions. Who are we? Where do we come from? What is this that we're living in this construct? Is there a big bang? Yeah, the Big Bang is when we hit the power button and turn you on. That's the big bang. That's when everything went out. Yeah. So you know, they can then maybe even begin to write programs in the Sam's program to create their own universe. So there's multiple layers of reality. I don't think we're even close to base reality, isn't it? It's

so crazy that it was only 30 years ago that we only had that Pong game when the two ball the balls and the two sticks. Oh my God. Now look, we have 30 years crazy.

I got the original Atari 2600 My uncle bought from me. He came down from New York. He was the police officer in New York, the only most expensive gift ever got, which is why I still have it today in a closet in a box. He bought that thing. And I played that thing. The Pong. And Pac Man. Yeah. And I played it for a couple of weeks. And I got the little callus in my hand. And I was like, Okay, I'm sorry. But it was cool. You know? Yeah, connect to

a VHF TV. And then look where we are. Now. My son turned a video game on this was a couple years ago. I was at his apartment. And I thought it was an actual NFL game. I didn't know what it was. Yeah, video game. Yes,

I play. I play video games all the time. And I play Madden and NBA and these games literally look like you haven't seen those games before. They look just like the real game on TV. Right? indistinguishable. Wow. And then now with like, with VR, like the Oculus and the stuff that Facebook's doing, it's pretty wild and augmented reality. Yeah. If you Yeah, if you like Elon says, if you assume any rate of advancement, compounding over time, you know, just in 100 years, there's going to be let in less than 100 years, as long as it begins, you're gonna be completely with 100% indistinguishable from reality.

Yes, that's true. And so the human mind so easily fooled. It just takes program input. So the human mind is encased in darkness. It doesn't even know what's going on out here. But it has friends, sight, hearing, smell, touch, feel all those five senses, it sends the friends out the friends collect data. Now, my friends that my brains, friends touch, doesn't know what this is, doesn't know what's going on here. But it combines with the data from the eye the eye sees, it doesn't know what it is. But the eye ended, and the hands send a signal back to the brain send data packets back to the brain, the brain then sorts it out, and then creates a hologram of what's here.

And then I'm navigating through the third dimension based on a hologram from data packets, right?

And the universe is doing the same thing. There's one consciousness, which has divided itself into Google's of entities manual, one of them. And it's experienced itself, we have a sensory perception for the UE have,

we only have eyeballs, nose, mouth, skin, we can only are constructed, physical bodies can only observe so many dimensions. Well wonder what it would take for us to be able to observe a different dimension? Well, we what are we thinking about it the right way? Would it be a physical alteration to our body or what? What

M   C   M   A   N   U   S

would be there

would be scary

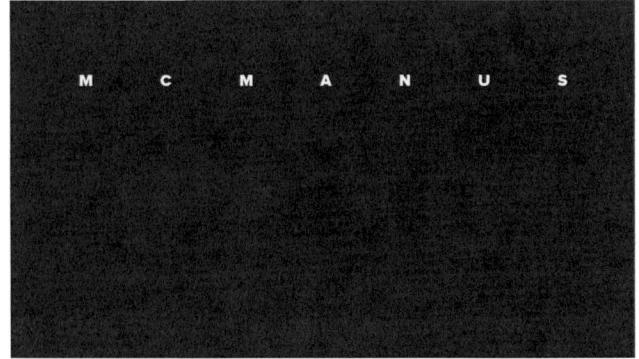

because we can't even see, we only see 1% or light spectrum here. If we saw more than 1% of the light spectrum, we'd freak out.

There's so much going on all the under multispectral and the ultraviolet and the infrared, and the gamma ray range and the X ray range, we would freak out this stuff going on all around us as we speak, billions of waves and frequencies moving around. So we will be able to see if we had the eyes for it. And I think that 1% is all we can handle. Yeah, more than 1% it would be a lot. It'd be cool to have some x ray vision or something like that. Yeah. But you add it all together at once. Now you're talking about multiple dimensions. Well, the human avatar body can't even get to the dimension ninja, because we're a third dimensional sub structure, we'd have to create the technology to face shift into a body of some type, or whatever type that would possibly be to even ascend to a higher dimension. Yeah.

Well, I mean, going back to the simulation theory, it seems like if we were in a simulation, of course, only like people like Elon Musk would believe that because if you were playing a character in a simulation, you would want to play somebody like Elon Musk.

Yeah, right.

You wouldn't want to play Joe Blow working at the hardware store. You people like with with fascinating, interesting lives doing crazy things with unlimited money, they could do anything they wanted, that would be the most enticing player to be in a simulation.

It's pretty rare. But what I think is happening here is the reason why it's not specifically like, again, you can you control it in a certain level. But what's cool, I think, is that the universe itself is using us to figure out what it's like to be Billy Carson, what it's like to be what it's like to be this microphone, what it's like to be a person that's living in poverty or sort of like to be a person that's living, which was like being an adventurer, and explorer.

What does it like to be all these different things? What is it like to be a blade of grass? What is like to be a rock?

Did

the US government make a treaty with aliens to allow them to abduct citizens, according to behold a pale horse by William Cooper. They did. 34th President Dwight D. Eisenhower allegedly did this on vacation in Palm Springs. So according to the book in 1953, astronomers notice strange craft orbiting the equator and through radio contact, they weren't able to set up a face to face meeting with the beings on those ships. The movie Close Encounters of the Third Kind is a fictionalized version of it. Meanwhile, arrays of humanoid aliens landed in the homestead Air Force Base in Florida and said whatever you do, do not make a deal with those aliens that are circling the equator. Now there's arrays of humanoid aliens, we would like to help you with your spiritual development. And they said we won't be giving you any technology because you're not even spiritually evolved enough to handle the technology you already have. And these humanoid aliens that the only way we're going to help you is if you agree to disarm all nuclear weapons and the US government was like, oh, man, we're gonna work with those aliens that you say are evil because they're gonna give us weapons.

On December 13 1944, the chief and Commander Dwight Eisenhower wrote up a press

release, which became this New York Times article claiming that mysterious floating mystery spheres might have been a German weapon.

How can I afford nails

the video burgers video which is a close up of a craft. As you can see the occupants a

conspiracy theorists breaks down some strange bad man posts from actor Matthew Perry.

So your frequency determines your reality. But how do we know what frequency we're at is actually this quote by Albert Einstein as well that says everything is energy and that's all there is. to it, match the frequency of what reality you want. And you cannot help but get that reality. And you can be no other way. This is not a philosophy. This is physics. So we have to be a vibrational match to that which we want to manifest in order to experience it in this 3d reality. And inside this book, we can look at the themes of consciousness, which are the different emotional states that we can experience from lowest vibrating frequency to highest vibrating frequency. And we can actually learn inside of the inner work, how to

identify which emotional states were consistently experiencing and stuck in, and how to transcend those emotional states to raise our vibration of frequency to that that is a match vibrationally to our manifestations. So if you want to do the inner work to see what vibration, you are emitting, what frequency you are emitting, and actually transcend these vibrational states to make yourself a vibrational match to that of your manifestations, then the first step is literally as easy as hitting that little button right down there. And get yourself the workbook on the TIC tock shot. Otherwise, guys, if you like the vibe here, if you want to see me again, haven't already hit that. Plus, make sure you're doing so so you don't miss out on any other future videos I post. As always, guys, Much Love y'all

storytime on how I met my best friend the acceleration of things that are going on on your plate, huge

mistake in the relationship between humans and these other civilizations. When we detonate the first atomic bomb and the subsequent. Those go off. Everybody knows electromagnetic pulses. What they don't know is that there's an attending so called scalar, or longitudinal wave that is faster than the

speed of light that disrupts the fabric, and communication and propulsion.

You probably remember this video where a man is holding a supposed Egyptians tone and ancient Egyptian style that has superpowers. In this video, you can see that his hand becomes transparent. And this tone does not reflect on the mirror. There are 1000s of videos like this one and no one knows if they're real or not. This video, for instance, something very similar happens. Check that out. Well, I've shown you guys these videos in this video, for instance, you can actually see it a little bit better. Right? Well, there's this other video certainly the internet right now. And well, I'm not sure what's going on here. Check this out

pay

attention

look at what's going to happen.

What is going on here? Hey, you're getting older, when you

I've heard from people in the Pentagon, that the buzzword in the secret of secrets in the Pentagon is the Sumerian gods are returning. And that's what they're referring to that whole area.

You repeat that again, just in case anybody missed it.

Last word in the Pentagon and the military circles that are in the know about what to cover up here. They kind of whisper tones talk about the return of the Sumerian gods. And they're talking about the what we would call aliens or fallen angels and returning into the Middle East, and tools or areas, the Anunnaki.

When you're trying to do telekinesis, it's not what you think you're not sitting there like trying to do something, what you do is you imagine a biofield over what you're trying to move. And once you can solidify the biofield over what you're trying to move you feel for it. And once you feel it, you'll feel like something pressing your hands. And from there, you can actually feel something and move it. And when you move the biofield whatever's in the biofield moves with it. I figured that out on my own. I was just diving into that. And as I was doing it like I can feel the biofield is the same thing with Reiki. So like even with my son, he was in a dead sleep. And I put my hand on him once I could feel his biofield I rested my hand on it and I kid you not he twitched so quick. And my girlfriend was like What the eff was that I touched his biofield. And the thing is, I can move my hand over his head like this and not touch anything. It's when I give awareness and I acknowledge his biofield and I try to perceive it with my hand and that's when I can sense it and touch it. It's weird how it works when you're if you seriously want a deeper understanding of what we are, this book will explain so much to you. Like I can't even begin to express how deep this book goes and what it will do for you. You have to understand that the truth lies Between all the religions and this book literally explains all of that. It's literally 121 pages and it just goes over all of it

it's just man, I'm still going through it and I can't even express to you the amount of information that is in this

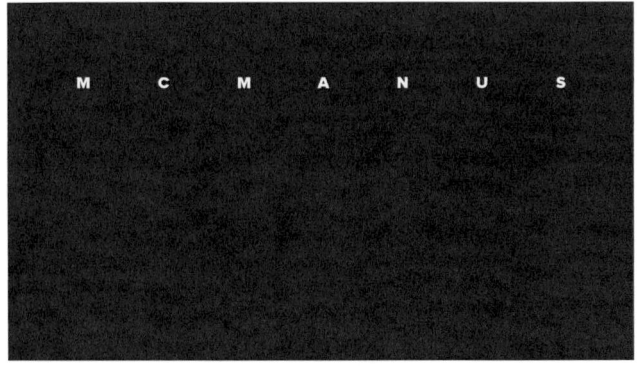

so there's seven colors of the rainbow there seven chakra system seven gaseous layers. There are seven musical notes Tilray me Faso Lottie and then it looks back around to do the point is that we have seven metals that are produced by seven luminaries. So on the bottom here you have the earth and at the top you have the ether, you have the luminiferous ether, which carries lumen like Lucifer holds light, Christopher holds Christ, the luminiferous holds the ether seven luminaries, the seven chakras as within so without, everything is energy everything is spiritual matter, the light of the sun becomes gold, which is why gold is worth so much Venus becomes copper later Mercury becomes Mercury, the moon becomes silver, Mars becomes iron, iron is in your blood, Mars's red, light of Jupiter becomes 10, Saturn becomes lead. Those are the seven different metals. Low frequency light becomes matter, which is angelic beings falling in consciousness, high frequency matter becomes light. So when you raise your vibration, you become illuminated, you raise your vibrational frequency to become angelic. And you do that with love. Your body is a temple, you raise your vibrational frequency through love. And that is how you get into the Celestial court. That is how you get into

heaven. And before you come at me with your religious or scientific views, this is just based off of syncretism, that of what I know.

is that the Earth is a closed system. We cannot leave the Earth. There's no place to go.

So there's seven colors. This is literally a strike and our freedoms, I have an account learning. So that was posted in October of 2022. It's November 23 2023. So a little over a year and they finally me, but I have and I got the hiccups. But I'm just like looking back at all my old videos and I'm like so chiseled. I'm gonna I'm honestly happy I fell off because I get to read document my frugal living eating nothing but raw fruits and honey, no protein, and you guys are just gonna see like the craziest transformation ever. But this is your electromagnetic field. This field has seven energy senses and they get done astral projection come natural without train. Yeah, if you can sit in elements of stillness for long periods of times, following my lucid meditation, it happens do my lucid meditation every day for three months, and I promise you you will accidentally leave your body and ash

I came across a document today from the CIA's escape. They

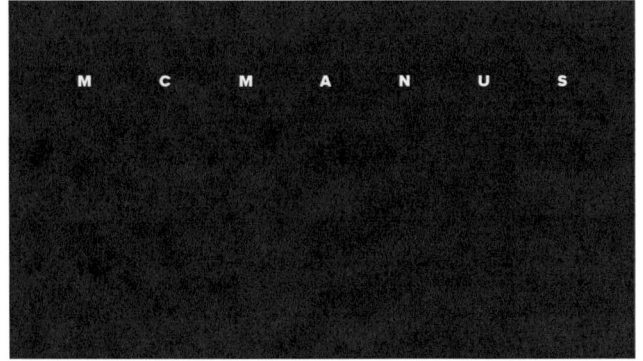

know you won't believe this. Have you ever wondered? What really happened to Malaysia Airlines Flight 370. Listen, as Dick Gregory explains, what he believes really happened to

for all other artists is that we'll know. Yes, you can get paid without a label.

Did you know that in 1998, this man Jeff leecher took a photograph of a real life mermaid. It was the 12th of April 1998. And Jack was on a boat about 20 minutes of chi we point in Hawaii. On this particular day, there were 10 people on the boat and a pod of dolphins appeared behind the boat. They were chasing the boat playing jumping out of the water. When all of a sudden one of the men on the board started shouting and pointing in an absolute panic. When Jeff went to see what this man was shouting and pointing out, he realized there was a woman swimming with the dolphins. According to Jeff, this woman was only 10 feet away, he could clearly see her swimming alongside the dolphins and keeping pace with them. Jeff went on to say that there was no way any human being could be able to keep up with the speed of these dolphins. And then she leapt into the air and my heart almost gave out on me. The entire lower half of her was covered with scales and tapered back into a huge fish tail. And then she jumped once more and disappeared. But that is not the end of the story. About an hour later, they arrived at their dive spot. And Jeff went into the water with his underwater camera to take pictures of the marine life and he secretly hoped you might see the mermaid again and be able to snap a picture of her shore enough during his die that he felt something brushed against him. And when he turned see what it was there, she was darting off through the water. He started snapping pictures and guess what he was able to capture a picture of the mermaid. There she is in all of her mermaid glory, a real life area, if you will. So according to this article, weekly world news sent the picture is to be analyzed by three noted photography labs, who those three noted photography labs, where I do not know what all of them said the photos were genuine and had not been tampered with in any way. So the general consensus apparently was that these photos were authentic. And I'm just gonna say it I think it's really spectacular that in all of her speeding about in the ocean, somehow managed to capture this incredible picture. And when you do some research into what mermaids would have to look like to be able to survive in the sea. It's less, half woman, half fish, with beautiful long hair and more. No arms and either lots of blubber, or for like a sea otter, or hair to sea otters of hair. I think it's hair. It's a good story, but isn't just that. Now Jeff maintains that there had been sightings of mermaids in that area for years and years and years. So who knows? Maybe what he saw was actually a beautiful mermaid from the depths of the ocean.

You do not want to scroll past this video I'm telling you right now, this is my last post on that movement. But it's very important what you're about to hear. At the end of that tumbling movie monsters of California there are coordinates given.

Okay, so every time that I mentioned Dr. Steven Greer, my comments section becomes a battleground. Whether you're a supporter or a critic of his work, the number one question that always comes up, does he truly believe that there's no such thing as a hostile race of aliens? Oh, piece of advice. There was a time where Yeah,

perhaps 90 million years ago, there was a lush rainforest that is now covered in miles of ice on the Antarctic continent, a friend of ours,

who will remain nameless went there. And he said a couple interesting things. He said the mountains looks eerily pointy and foresighted. He said way more mountainous than he thought it would be. He said there is a pack with every country that has a slice of it that they will not dig or remove any minerals from Antarctica they are clearly protecting or preserving something that may or may not be under I believe it's three miles at its thickest of ice.

Some of you might know me because of a story I broke last March concerning something called the guardians of a lookingglass. Basically, what we had here is a technology

That was built, which would give those people who operated that technology, of course a huge advantage because they could see and view timelines within a stream within a field. The awareness of this technology came to us through a certain whistleblowers, and one of them was a guy by the name of Dan borsch who went into some detail about how that technology was acquired starts at the end of the Second World War, it was the first time that the United States ever came into contact with certain ETS, you could say, who began appearing over the skies of the United States, and their appearance game was accompanied by crashes and captures both the craft and the pilots piloting that craft, and it leads directly to the formation of the breakaway civilization, that which was created by one ultra ultra secret society, that was called majestic 12. They have ultimately come into possession of the looking glass material, and time travel and other technologies ever since. So between 1947 In the early 50s, there were a series of crashes that took place, not just one. And they started the whole back engineering process, and they also had captured extraterrestrial beings. And the Looking Glass has ended up in area and yes, for area 51.

The program out at area s four consisted

of three projects, Project Galileo Project, sidekick, and project Looking Glass project Galileo dealt with gravity propulsion project sidekick dealt with a beam weapon that had a neutron source and was focused by gravity lens project Looking Glass dealt with the physics of seeing back in time,

what they really discovered, in principle is that the ETs are not really truly ETS their us from the future. They call them J rods,

a long period of introduction into needing, you know, say, a Navy. I call them J rod, of course, that's what they call them. But that wasn't his, you know, I don't know if that was his real name or not, that's the name that The Linguist gave

came from Zeta reticular, which is what we are going to colonize in the future, and they traveled back 52,040 5000 years respectively.

If you live in America, I'm about to show you something very shocking. And quite frankly, every single one of us Americans is filming this right now, especially if you have a family. But this video right here will let you know what time we are living in, and how bad things have gotten. The cost of living is through the roof. And here's why. Check this out.

I've seen two videos in the last week talking about the same thing. And they're on point that they're forgetting one big fact. So I'm going to cover this right now. This is what a lot of people don't realize almost everybody I know realizes that life is more expensive now. And then all of us are being outpaced by the cost of living. The cost of living has been sky-rocketing for the last 20 years at least. And the you know, the rate of pay and salaries, you know, basically our compensation has not been keeping up with not even close. So all of us know this. Back in 2006 2007 2008. We went through that whole thing. And I remember how crazy the inflation felt. And I remember seeing a speech by Ron Paul talking about how it's not really the goods getting more expensive. It's just it takes more of our dollars because our money is becoming less valuable. That is true. All right. Now a little fact, here's here's a fact you may not know maybe you didn't know this. The Great Depression. Okay, if you go back to 1930 1933, that period, it was considered to be economically the worst time in the United States history. It was the time when everybody struggled, people were eating rodents on the street. And if you go and do your math, you figure out that during the Great Depression 1933 I think was the worst year. If you take what the average income 25% of the workforce was out of work, the unemployment rate 25% is pretty high. But the people were employed, the average income was between three and $4,000. That's a real number. Now, if you if you take that number three or $4,000, go with whichever number you want. And you adjust that for inflation and cost of living to today's money. And it comes in right around between 80 1090 $3,000. Back in the Great Depression, people were struggling and starving, because they were only making 80 or $90,000. But really let that sink in for a second. So, now those two videos that I saw, I've said that, okay, and that's a real thing. I get it. One of them said, like 3300 bucks, the other guy said four grand. But anyway, it's between three and $4,000. But here's what they forgot to mention. You don't they didn't have a 1933. All of the taxes and fees that have been piled on top of us over all these years, more than 70 years. You guys know what I'm talking about? Okay, we've got 1000s of taxes that we're paying now. We've got gas tax, cigarette tax, Sodapop, tax, you name it, they didn't have that back then. If you just look at statistics, you're going to also find out that they added on income tax and payroll tax, those are two new giant taxes, the taxes skyrocketed right after the Great

Depression. And it compounded by three once payroll tax and income tax that added to that. So they basically tripled our tax. And ever since then, they've been printing money like crazy, which is making all of our money less valuable, which is causing a bigger problem. Okay. So not only would we need 80 or $90,000 a year right now, in order to match what people were struggling with during the Great Depression. But right now, we would probably need double or triple that to make up for all the taxes that we're paying. Okay. All right. Now, it gets even crazier, because if you guys haven't noticed a trend, and I'm going to end on this note, there's a little stupid trick that our government is doing now, and you guys should be aware of it, we should be paying attention to it. And we should be screaming about this. The government has figured out that, okay, taxes are hard. Why? Because no taxation without representation, we have to vote on an approved taxes, you don't We don't have to vote on or approve fees. And they have figured out that if they just call something a fee, it's not a tax, and they can just make it up and charge it to us stop and think about all of the fees that you're being forced to pay. Now I'm going to call bullshit on that. A fee is something that is a choice. I don't think that there should be any mandatory fees because a mandatory fee is attacks. So I'm just gonna call bullshit on that. And I think awareness is the first step I think all of us need to be aware of what's happening. And speaking out against this is going to be the next step taken are kicking our government to task on this stuff. The transparency needs to be there and accountability needs to be there. And these politicians that are getting elected that aren't actually doing anything to fix this problem. They need to be held accountable, that I'm gonna end on that note for this video.

If you've been iffy, or you haven't gotten this book yet, because you're like, Okay, well, is it worth it for the price? I don't want to spend that money and then not like it or not work for me, this is your time. Right now. They're having a fall sale for these books. This is originally $21. It is $4 Right now $4 $4. That's literally insane. You cannot get a better deal than that. So if you're someone like me, where it's like, too much of a price to even see, but we weren't the premier, not this, like I said is your opportunity for dollars, what do you have to lose? You have $4 to lose, but that's not that big of a deal. So I will say though, for some reason, Tiktok is kind of weird. So it's going to be $4 for some people. And then for other

people, it might be like 16, which is a big jump when it comes to price. I'm not really sure like who gets what, but I do know majority people get $4 for this instead of 16. See what it is for you. And if you want to try it, try it. It's very worth it.

So I

found this in a cave. This thing is 2000 years old. And one of these pages it really focuses on the fallen angels. And this blows my mind every single time because this is called no joke. The Book of Giants.

In this book, it talks about how the angels bring both knowledge and havoc the angels exploit the fruitfulness of the earth. Now when I hear this, I think they were building empires. I mean, could you imagine a city built by angels? But that's not even the most interesting thing because there is this one sentence. It talks about the angels taking 200 animals from all species and genetically modifying them not kidding. In other words, it says right here from every Animal from every bird for miscegenation. miscegenation means cross-breeding, it's genetics. It's the highest form of science. And it kind of makes sense. Actually, the angels were making their own creations

they were playing God. And that just begs one question, what would a fallen angel ever want to make? Right now

I'm going to show you how to figure out if someone's an NPC or even if you're an NPC real simple, take your hands together like this. Together, move them apart, slowly together, but don't touch. If you have a soul, which the sole source connect energy, emotional energy, which is transmitted throughout your whole body, if you have a soul, you should be able to feel some energy.

And now, you know, right now I'm going to show it.

So

you all wanted me to talk about time maps.
So let's do it. When

I was called on the telephone, and I was told to go to a certain address, I just got in the car and went there. And I'm gonna tell you all something that walked into with looked like a normal building. But when I got inside this building, I never seen like, anything like that inside, because all along the top of the building on the upper rows, it seemed like invention after advancement after invention after invention. And it went on and on and on and shelf are things like magnets that picked up gold coins, gasoline that was made from plants green, which he called Green gasoline coils that I've never seen, almost like a Tesla COVID Much deeper angles, which were all resonate at a certain frequency, pay very close attention to one thing, there was this chart. And it was basically laid out like this here on the chart, but it was huge. I started asking for this chart, he said, well go there Look, I went over there. And it showed where time started on our timescale, and then went on and on and on. And he got up into somewhere in here. 2001 was about here, I was in 84, I was back here somewhere. And there were little pictures underneath the chart. And I know that most people are going to find this hard to believe but all along this chart had said who the presidents were, whether they were assassinated or not assassinated, what the year was gonna bring as far as his food for the world. And when I got up here, said that the world was in a chaotic condition, basically, in 2000. And it showed more in Warren, Warren, Warren, and more Warren, and also showed that name, the President's by name, it showed Reagan it showed Carter it showed Reagan and showed George Bush Senior. And after that it showed Clinton is in the Clinton years, right. And I was back here. How they knew he were the President's worth was a mystery to me. And when I asked him that question, he says, We know.

in this video, I'm gonna go over four people that you can turn into a puppet, when it comes to your spirituality. Number one, your significant other, either your husband, your wife, your girlfriend, your boyfriend, your fiance, whatever you want to call it, having a puppet with us, them is to control them. So you know what I mean? So they can do, as you say, you see what I'm saying? I know people disagree with this, but a for some of you that want one and want to control your relationship, having a puppet of them will benefit you. Number two, the judge, if you haven't court issues, you got to go to court all the time, you want to have a puppet under the name of that particular judge. Number three, when it comes to your manager, your supervisor, or whoever it is at your job to make decisions above you, you want to have a puppet so you can control them as well. And number four, your landlord save giving you problems, turn them out of pocket into a puppet. So now those are the four people you should turn into puppets when it comes to your spirituality. Now you want to know how you turn them into a puppet. That's a different video that will come pretty soon but for right now, this are the four people you need to have right here.

There's this there's this to this. The ego mind of fear is extremely suspicious, protective, defensive and insecure. Its core beliefs are rooted in seeking security and the illusionary promise of safety from loss In order to attain that safety it believes being anxious, paranoid and doubtful will help in its mild expression. It comes out in the form of social anxiety, fear of being lied to fear of being abandoned fear of being rejected fear of failure, fear of loss, fear of getting sick, fear of injury or fear of death, as an overcompensation to the fear of survival. We may overly expose ourselves to things that invoke fear such as constantly watching frightening movies, being a daredevil, a survivalist, or a thrill seeker always tempting death, the ultimate fear, the adrenaline of fear can become highly addictive. And as it is continuously experienced, the brain becomes desensitized, thus pushing us to seek more dangerous and scary experiences to get the same rush. Inadvertently, we become addicted to stress. So I took the initiative to do a deep dive on a tic tac shop to find an alternative because apparently everybody's afraid of shadows and mirrors.

And I found this book called The inner work, it is a book it is not a journal. And first and foremost, if I bring you a book like this maybe I did not read it, I studied it. And when I tell you, this is the chef's kiss of healing and working on your wounds, chef's kiss. And what I found so interesting about the overview

of

fear, or the emotion that's attached to fear is that we talk about addictions all the time, why we tap addictions to sex, men, women, clothes, food, shopping, we have all these isms of life that we never ever, ever address the fact that wounds that are associated with being overwhelmed, and adequate loss. All of those other isms as well keeps us addicted to the adrenaline of stress. We don't talk about that. This isn't had to be more than a part one video, baby, I may have to go live with this. I have $400,000 my account now because I've been making $300 real cash every day,

being led into a room emphasize had been watching a television transmission from Chicago.

And policemen had just shot a black man that was running away. And then this is

African voodoo witchcraft. And that thing is called battle or the spirit of night. There is nobody inside of it is actually a spirit, which is a guardian of the Black Magic community in Africa.

This is a group of beings and the ancient texts called the EGCG and the GG word, the working class version. But they weren't supposed to be slaves, but they were being treated like they were slaves. They got tired of this work and decided to go to war. So they actually got together. And they said, We're gonna go against God. I knew and his two sons and Ian and Lille and so what happens is he says, You know what, I have an idea. Let's not go to war. There's an existing being on this planet, we can actually modify it and add our essence to it and get it to do the work for us. That is when the genetic modification, you're looking at investing money?

Well, well, well, once we have the flags of all nations. I wonder if there's any interesting ones?

to deduce precisely. Study this American flag with 40 Star 40. No jet GBT. When did the US have 40 states, the US had 40 states from 1889 to 1890. So we know that the document is between the dates of 18 and 1890. And the pirates have their own flag. Super cool. That this Chinese flag looks unique. I wonder what time period flag of the Qing Dynasty was adopted in the late 80s. This is the closest you're gonna get to a real picture

of a

UFO.

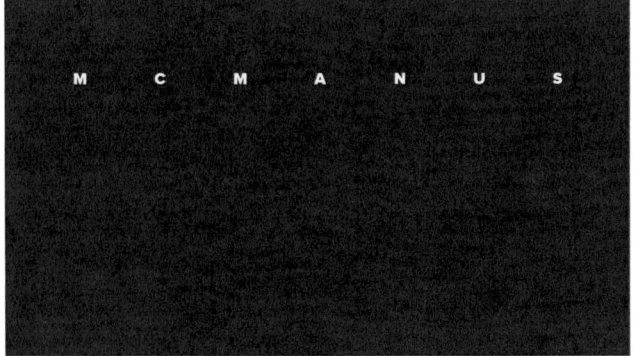

I didn't nobody told me about telling my mom that you guys wanted to hear stories about witches and she left me to voice messages and I listened to them today. It's not funny, but y'all listen to this. I'm gonna play it in Creole just a little bit and then I'm gonna translate it for you guys in English. You can see

right away, you're making these cute little skull earrings. She said that back in the day. When they used to go to church. There was this which one morning?

They came to church on Sunday. And the pastor and now a really good genuine pastor with real serious power in Haiti. I can't speak for anyone else. But I know in Haiti. They be some real godly people. The pastor goes I know that there are some People who come to church that are witches. On Saturday, y'all fly and y'all go do all of these evil acts and then on Sunday, y'all come to church trying to pretend like y'all righteous like you're good people. I know exactly what you are, and I can see through you. And I'm warning you right now, when it comes time to receive Communion. If you come in and you get communion, I promise you that it will be stuck in your throat. No, it is right. People just be like, they bluffing. They can in no way. No way. You know, the time comes, it's time to get up to receive Communion. Why this lady know that she's a witch or Lugo. And she really walked up and got communion. Oh, sure enough, just like the pastor said, the skin thing got stuck in her throat, and she couldn't talk anymore. She literally became mute. So then her daughter comes back and remember, which is most of the time they will have children. Her daughter comes with her. And her daughter says to the pastor, like, can you please undo whatever it is that you did to my mother, she has been mute. She can't talk ever since she went up to get communion. She's not able to verbally communicate. The pastor just looked at her and said you knew what you were. And I gave you a fair warning. I told you that if you come up and receive Communion, what was gonna happen to you, and you didn't heed my warning, what is done is done and it cannot be undone. now deceased, this woman spent the rest of her life mute. I told my mom that you guys want to say this is

one of the most scientifically researched natural supplements available to you. black seed oil was once referred to as panacea, which literally means answer to all problems. So let's get into all the benefits. It's an anti diabetic, it

lowers blood sugar and improves insulin sensitivity. It's a pain reliever and it's anti microbial. So that means it fights off bacteria, viruses and parasites. It releases muscle spasms, and it's full of antioxidants and anti inflammatories. It's great for the lungs, diarrhea and acid reflux, liver disease, asthma, allergies, arthritis is good if you have high blood pressure, and it improves memory and cognition. It's good for eczema. And if you want to improve your skin in general, it's going to make your hair grow longer and thicker. In Maine, this is going to be really good for our testosterone, women, this will stimulate breast milk production. And it's also great for your good, which is your second brain, it fights off the bad bacteria and most important good bacteria, bloating, constipation, indigestion black seed oil is going to help with all of that it's good for literally everything. When I read the benefits I was mind blown is so excited to know I need to really take control of my health and put life and good energy into my body. Okay, too. So this is

if you can understand what I'm about to say, thank you for the work you're doing. There is a shift in human consciousness that is currently happening, pushing humanity to disconnect

and look within so that we can truly understand ourselves, so that we can better understand the Mueller world for the purpose of evolution, an event that impacts our entire universe. See, we cannot heal in the same place that hurt us. For now, this shift can only be felt by those who have awakened. And it's been described through many names ascension to fifth dimensional consciousness, the return of Christ consciousness, spiritual awakening, among other names. But all of those are just labels for that just one core happening, which is consciousness. What is consciousness?

From my personal experience, when I first started experiencing this shift, it was like I finally became aware of myself, but not the identity self, the self beyond this physical, temporary body, the observer, I all of a sudden realized that what I thought to be true, was not it. I had no other choice but to surrender. Everything that I thought I knew, which in truth were just programs, beliefs, concepts and ideas that I have picked up along the way from others. While I was looking for myself outside, I now know that I am everything. But I am also the nothing that becomes aware of itself. I can now see the interconnectedness of everything and everyone. This is how I now know that coincidence doesn't exist. It's synchronicities of an invisible network that connects all things. And everyone. Albert Einstein said it best. He said, coincidence is just God's way of remaining anonymous. Now, in the coding of the Bible, and key word coding. John three, seven, you must be born again. This is it. I felt that the person that I was had died, and I was experiencing a rebirth. And at first, I thought that I was going crazy. But knowing what I know now, I was actually becoming sane. In a world where insanity is considered being normal, realistic and falling in line with the person in front of me. Yeah, no thanks. I'll think for myself and I will carve my own path. See, during this time and where you're being is experiencing this rebirth, your mind truly becomes that which resembles the mind of a baby. You are now learning and observing new knowledge through frequency and vibration, which is how the mind of a baby learns in its early stages, you become aware of all energy, which can become overwhelming. Your mind becomes a receiver for new information information that is now coming from within. In addition to the knowledge and experience that you have gathered throughout your life from your exterior world, thus it becomes a

lonely path. You can now see through everything, and everyone. The barriers or curates
are simply an attempt to respect the superficial privacy of others. This allows me to use
Bayesian interference, which is a method of
generating predictions based on new information experience, prior knowledge and belief. This is where spiritual discernment devel-
ops, as you are now operating from a state of
knowing. Although belief is what creates this
three dimensional reality, belief ceases to exist in this higher state of consciousness. It's
I describe it as direct access to God, which
was once just an illusion. But once you truly
get to see what God the universe is, which is
all within you, this world then becomes the
illusion. Awakened souls are now taking the
lead to show others with the yet Canarsie, it's
acting as a wave, which will eventually realize it was always the whole ocean. I call this
first wave of awakened souls, the ark, cou-
rageous and strong to take the ridicule the
judgment. Let me remind you that all great
visionaries, the greatest minds, the greatest leaders to ever walk on this earth, were
first ridiculed, humiliated and condemned by
those who couldn't see what they could. It
is only now that their visions have become
our reality, that we praise them. There is no
stopping this sort of thing consciousness, it is

MCMANUS

our future, we must match the frequency of where our world is headed. A beautiful world of unconditional love and unity, a world led by the heart in balance with the mind this

is how you get a free

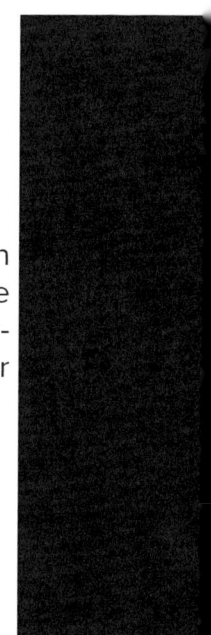

kind of alien or into your you have a When do they come visit you? Like I probably like three in the morning, I think you go any-where yet, I don't really want to do on your planet

The goal is

these are only $1 Tic Tac shop and I'm about to show you guys how powerful these little guys actually off and why you should probably spend $1 and grab first things first. They got to

this might be the only okay so like I said in
my last

video I was there is an herb that can eat and devour witchcraft that is sent towards you no matter how bad it is just like a mushroom eats plastic. It's called the evil oboe you We'll be thankful when people send you curses because it actually turns into something that's good for you. And it takes everything that person has away from them leaving them, there is an herb.

They tell you not to stare at the sun because the sun activates your brain as the only reason why after staring at it, you close your eyes, you still see light, it's because that lights entered your brain and it's activating it. You're not supposed to be a black mindless fucking drone. You must be a light, a creative beautiful intelligent being. But those types of beings can't be controlled.

What, Marina is the fifth dimension

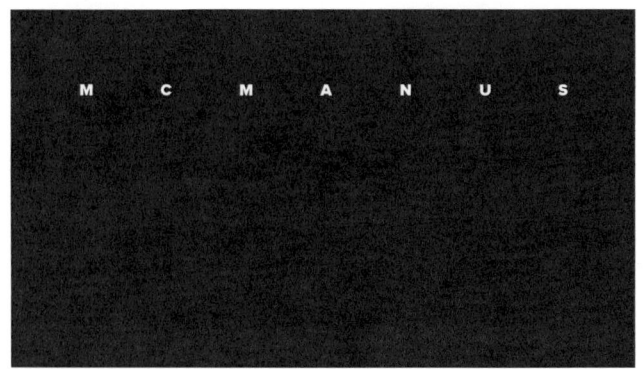

the fifth dimension is created by particles itself that are co creating by reprinting the structure itself one on one, one on one plus equals three creates a certain structure of modality that is reprinting on the quantum field next to itself next with itself to infinity. So basically from one particle you creating a structure that is expanded to infinite and collapsing to infinite that collapsing can go to infinite realities of reprinting and expansion of infinite reality of reprinting. If you create the modality of a similar structure, some vibrations, they're going to reprint in the similar timelines of realities encoding parallel timelines

that was directly channeled.

that was directly channeled.

Have you had this ability that when you tap in

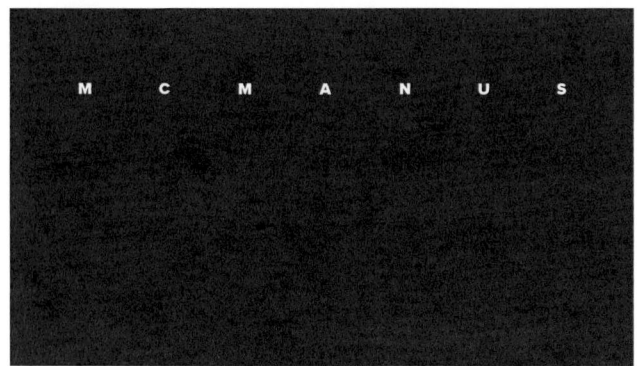

ladies and gents, make sure that you watch this video until the end? Because this is important for both. But ladies, I'm getting a lot of messages people asking me, you know, women asking me what is a good way to cheat Trump man, and one of the ways to trouble man is through your menstrual. I know this sounds disgusting. I know many people are gonna say Yo, this isn't a rose. But this is what it is. Many women that are serving their menstrual cups when it comes to oil. And when they are actually, you know, put it in red wine and put it into. So this is a good way for you to trap that man over half the men not leave you forever. I know many situations that you want us to help you with that. That's it. And also, fellas, be careful where you eat. Don't eat from nobody, you know what I'm saying do not eat for no women because you don't know what their intentions really are. And we're in the age that everything is possible. So just be careful and just watch out.

If you stare at your reflection long enough in the mirror, it'll start to stare back at you. What I mean is it'll literally move different than you. And it might even start a conversation. I'm not exaggerating, this isn't a metaphor or a joke. The mystery of the magic mirror is a secret that's been long hidden by the secret societies. Some of these societies refer to the ritual simply as memento mori, which means remember death. This exercise isn't for everyone. But I promise you, if you stick with it, your mind will be blown up and you will have my first key to the apocalypse. I'm going to tell you how to do this practice of first I want to give you a warning and an explanation. First off, when your reflection starts to do things that you're not doing, and even starts a conversation with you. I want you to ask yourself, Who is that? Is it God? Is it the devil? Is it you? Or is it the universe itself? The key to understanding what this place actually is lies and understanding what starts to talk to you and move differently than you

in the mirror. This experience

is going to change your life

by shattering your foundations of under-standing of this reality. If successful, it will unglue you from your perceptions and open you up to the real world. You may even begin to Dream Awake. I just gave you a hint. So here's my warning. Many people who have done this have had their consciousness shat-tered and it took a very long time for them to recover. So you perform this ritual at your own risk late at night, preferably alone. I want you to go into a room with a mirror. Turn off all the lights but get yourself a candle. Light the candle and put it to your left side. Make sure the candle is positioned in such a way that both sides of your face are dimly lit. I want you to stand about a foot away from this mir-ror and stare directly into your eyes or in the middle of your nose. I want you to start taking deep breaths and relaxing. don't anticipate anything happening. Just let yourself go. Be calm. Breathe, and let the process unfold. It may take a couple of attempts or may hap-pen the first time. First your face will disap-pear. Try not to focus on any given point. un-focus your eyes you'll see a shadow across your entire face. Then your face may morph In the various characters, I got a Tiger face first, then I turned into a caveman than a Grey Alien. This was very vivid and real. It was un-mistakable. When I pushed past the fear and amazement of that my face eventually returned to normal. This is when you have to wait a few minutes and be patient. The next thing, I will not tell you what happens. You'll have to see for yourself. Once you're done, be sure to come back to this video and comment what happened. This is the beginning of the realization of what this place actually is. This is a key to the apocalypse. You're welcome.

In the next few minutes, I'm going to show you how to open your third eye. your pineal gland, become aware of your true identity. This is a Kundalini meditation. The words you're going to be chanting are such Nam, essentially means truth is my identity and an easy pose. That's basically sitting cross legged. You're welcome to just sit in a regular chair, it's fine. Just make sure your feet are placed firmly on the ground. And then I want you to take your two fingers, your, I believe this is your Saturn in your Jupiter, you're going to tap right here on your Third Eye Center. Close your eyes and just tap gently. Tap, tap, tap. Do this for about 30 seconds, you're preparing your Third Eye Center. Put your hands in Jan mudra. That is the Saturn or the Jupiter finger that to your thumb rests those on your knees and keep your spine erect. Begin by taking deep, long breaths. Close your eyes and focus your eyes on your third eye center to think as if the light of the universe is coming through from your 13th chakra down into your crown crew into your heart energy coming from the ground, through your root center up and meeting at your heart getting to chat silently in your head Satnaam sat, you inhale Nam when you exhale. Remember that the light of God the divine light is shining through from the 13th chakra right into your crown, connecting to your heart gaez energies pulling up into your heart center. And they're meeting there and they're connecting and they are grounding you and expanding you so that your energy bodies are more expansive, do this for at least three minutes. Do this for 40 days, and you will know that you are God. And you will feel tapped in precious that your authority lies within you feel authentic. Also know who you are in 3d, and what your mission is. And the next few minutes. Let's say that last sentence again, everything that we see is a projection of what's happening within our own mind our own conscious and all of the conscious together. That's what God is. God is not someone off up in the sky that's preaching down looking down. Like you say in electrical, you're going to hell no, God is a collective consciousness of everyone. I'm God, Your God, everyone here is God.

Okay? If what you say is right, and I cannot perceive reality, I just make up reality in my head with my perception

doesn't mean you make it up. But you are able to interpret the reality differently, and how you might interpret it right now. It doesn't mean that it just made up. It's how you perceive things

early

within

Where does your rational mind come from?

My rational mind comes from my consciousness. What does that mean? My consciousness is basically how from birth if my life perceives all of the information around it, I'm sure you know the Trinity body, mind and spirit when we're born. We're born with the spirit, the new mind and the new body. So we have the collective consciousness, which is the spirit, which is in all of us. The body is the physical way that we experience the reality and the mind is the mind that interprets it.

The actual space travel has existed for 4.4 billion years. The interesting thing about our galaxy is the fact that we have many different races living here, in particular, the human race, which we are a member of, and what we call the reptilian races, which we are not a part of, but who very much don't like like the human race. Apparently this battle that's been going on has been going on for approximately 3.8 billion years. Between reptilian races and human races. Many of them planet many of this, of the races that exist in our galaxy migrated originally from Lyra during a war. Apparently, sometime in our ancient past a group of reptilians were exploring looking for minerals looking for food resources, they came across a star system that was full of human beings who were agricultural, in their nature, abundant food, just they just had everything together. And the reptilian races they're called the C car in the Andromeda Tang we know them or we would know them from as being known as the proponents because they're from Alpha proponents and apparently they as a race were brought to our galaxy fully formed

excuse me on this one. And excuse my low vibration energy on this one. Let me explain okay. I received a vision last night when the most high so y'all just sit with me. And I notice fear y'all

get rid of the fear Last night I had a vision of where I saw I saw Demons, demons. I saw the demons cheering. I saw a swarm of demons about a mile long. Even further further, I could see massive crowds and crowds of swarms of demons cheering

and often a distance I saw something they were carrying these demons were cheering. And they were carrying a giant casket above their head. You know? This casket? I still don't know who this is, because I don't. We're not allowed to see. But I saw the demons cheering y'all. And they were toting his casket. Eloy, I saw the DMOZ toting his casket above

their head. And they were cheering. This casket was the beast inside of this casket. These demons were pushing this casket, totaled it head to head, passing it along. Forward, I looked up, and I saw a portal, which to me appears to be CERN. Okay, this this is the closest thing I could think of was like sixes swirling, Portal was opening. I saw 18 on it. The 18 started spinning so fast that the eight turned into infinity and the line of the one turn into a circle around the eighth. As his Portal was opening, I saw the demons cheering and holding their king into our world. You understand what I'm telling you? At the same time, I saw the angels the angels were up here they were pacing back and forth yard it was picking up the armor they was it was ready for war. It was picking up the armor ready for war. And I saw the saints man we're not so I heard the saints saying this over and over Whoa. All

the angels were beating their chests. They're getting ready for his wall. It was getting ready to spill onto our physical rim.

Excuse my energy right now. But you know what I'm saying? This was a vivid one where I walked in this energy. Yeah, so

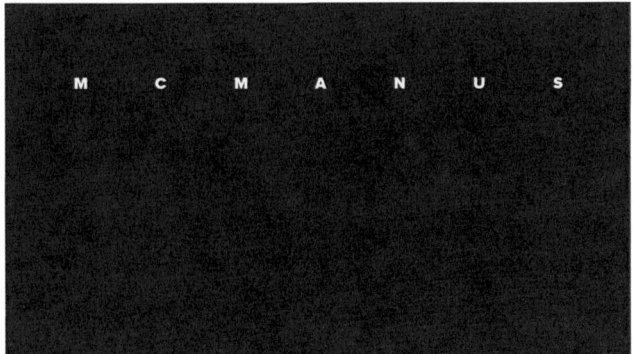

I have been getting this question a lot regarding what type of castor oil it needs to be. It does need to be raw and unfiltered. Now if you actually click this comment and go to see comment, this is the video where I'm talking about the chakras and that the third eye is actually associated with your eye health. So if you will have some castor oil on your eyes, before you go to sleep, it will start to improve your vision. And in a previous video where I was telling you out of stuff that I have on my house that I feel like everybody should have in the house. I was talking about it as well and people kept asking me, Where do I get the castor oil? Where do I get the castor oil. So this is the one that I use from the Tick Tock shop. Let me stop shaking it because in that last video y'all was ready to tussle with me because I'm saying the chalk can see baby Listen, this is what it looked like. I don't want no smoke, which I

am listen yo check this out. Yo, I just see I see something in the sky. Yo, I'm telling you. So

this is a video of

they'll be bringing lots of pyramids, planets past the outer. Now I think that y'all are not aware of certain shit. And see that's what do you believe that? Absolutely. Still here on the night? No, and tell me tell me I'm lying. You're lying. That's the problem. Where something you know. So tell me what happened when you were abducted. Before we move where I told you started like five times already. Okay, I haven't heard whatever my friends see my

These are completely inhuman creatures. These are either some kind of Fe like somebody said maybe some kind of demon or daemon related creature or something else entirely, but they are completely inhuman. They do. I don't know if shape shifting. I mean, they do make themselves look human. I don't know if they have another more natural form. So they could be called shapeshifters. I am unsure. I don't know if they can be conjured.

Yet people do mix up the rake with Skinwalker. But rake is a very, very modern term. In fact, there's not really any evidence that it was used before the early internet. I think that that's just a more that's just a more modern name for a changeling.

I will show you something that I could be a

fairy.

Their intentions are difficult to say what what they seem to try to do is get closer and closer to individuals usually alone. And the people that experienced them report an incredible dread that kind of drains them It feels like almost like their lifeforce is being fed on. So you could maybe consider them something like like energy vampires or something like that. This is a changeling a creature that tries to imitate humans, I also believe that these things are the reason that we experienced the uncanny valley. If you're unfamiliar with that term, that's the name for this strange dread and fear that humans get when they see something that looks very close to human but not quite right.

And a lot of people wonder why we feel that way. And I believe it's because of these things I don't know how they move when they say move through space.

They are quick there are kinda like they're not like super humanly fast, Slender Man I do believe is you know, like, that's just a story from the internet.

But it could be something similar the idea of

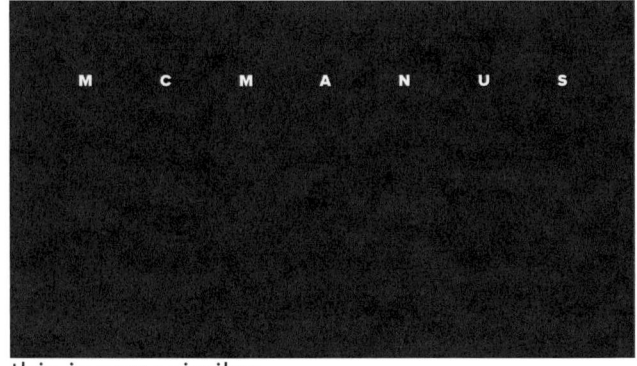

this is very similar.

Sorry, if I'm missing questions, guys, if I don't answer your question, you can go ahead and try to get it in a few times. I get a lot of different questions in here and the the chat starts moving quick.

I do not believe that all these creatures come from the same place but if you've been all changelings, I imagine Yes. I imagine Yes.

I don't know I'm I mean some of them come from a place you might call a different universe. But I'm not exactly sure

someone said good morning from Australia. Happy to have you there.

I don't know about cannibals turning into rakes. Someone asked me about about when to goes before I always mentioned that my information is primarily on creatures from Europe. I don't really have that much information about things indigenous to the United States.

Hello in Texas Ray Ray. Nice to see you

Jacob

Hello Jacob.

Okay. I do not know if they live forever

Kim thank you so much for the gift I really
appreciate the gift as you guys
are awesome

I would recommend you don't become a vampire

hello in Denmark always happy to get people from different places.

I tried to go live at different times so that people in different places who miss my previous lives can catch so hopefully I got a bunch of new guys here.

Thank you see bears you gifters are awe-some.

I love you guys. You helped me out a ton.

These creatures here were not in the US. This was a dungeon that was located in Europe. I do believe I know where it was. It is a ruin now, but I don't give that away.

Thank you Christine for the rose. I really appreciate you.

I don't give it away for reasons I'll get into as
we continue through the book.

Thank you, Carl, for the ghost. You're awe-some. I really appreciate you gifters you guys rule.

How did I become involved in learning this

thank

you

it was because I inherited this house that I'm living in now from my estranged grand-mother. And basically

Thank you sock monkey. Thank you.

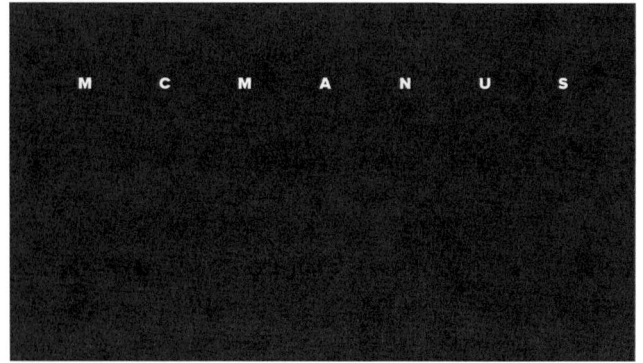

So in your case, I found all of this information within the house hidden underneath in in a trunk filled with books and tools and all kinds of stuff. I have the whole journey on my page. You can look you can look up when when we're done. You can check out the whole story. It's all available there.

Vampires are Real you would not love to become one it's not like the movies.

This thing here is what is referred to as the reanimated. You can almost think of this as the closest thing to Frankenstein. But it is still pretty different. These creatures are brought back to live

Thank you Steve.

Steven appreciate the gift are so much.

I do believe in aliens personally, but I don't have anything in here about aliens.

Thank you Alex.

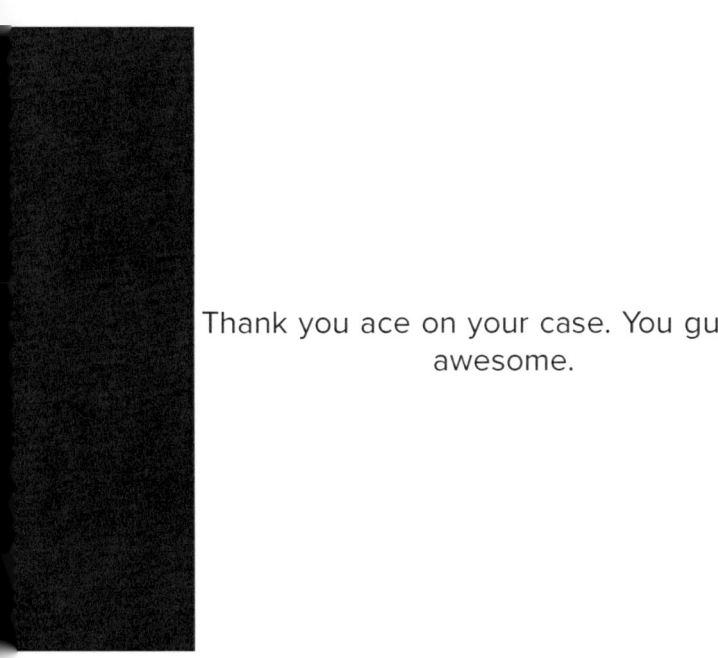

Thank you ace on your case. You guys are awesome.

MCMANUS

You gifters are just too like I appreciate you guys more than you can now. Okay, let me this the reanimate he is not wearing armor as I was saying these things are created through a combination of occult sorcery and crude technology.

And Alex, today's on your case, you guys are awesome.

Thank you,

Carl so much.

Thank you time is now the reanimated as I was saying there are a combination of different things. The idea is something like necromancy resurrecting the dead. But at this point in history, it seems that whoever was involved in this kind of stuff, was trying to streamline the process to use technology. Not only technology, obviously, but to try to make it easier.

This one here is I would consider a test subject. Because I have reports of things that are much more monstrous, much larger, multi limbed animal parts. Whereas this one seems like it could barely move and likely had no senses. This was a test just to see what could be done.

I always say this is the one thing that was held in the dungeon that I really, I really feel completely sorry, for. I do imagine that some of these things are still around today. But a lot of them were wiped out. I believe that people like my ancestors and the sons of man, were very good at their job that said you do you do still see videos and images and things of creatures.

You can find them here on tick tock. I've posted some videos going over more recent.

more recent sightings. But there are like I said, there are still some reports of like dog men of mermaids of rakes, and people say skinwalkers but they probably mean change-lings, things like that.

Someone's that?

Have they ever fought with humans things like this? Yes. I don't mean the Templar Order. It was called the sons of man. I don't know if they have relation with the Knights Templar. But that is the order within the Catholic Church that hunted these things that ran this dungeon. I have been contacted from people that I believe are the sons of man, I have videos going over it. But I have stopped taking blocked calls. So anybody who's trying to reach me who's not willing to be open, I don't I don't communicate with

don't know if these things were controlled by the people

that made them, they seem to have some level of independence. But you know, there probably was some stuff.

One thing that our generation did well, one thing that our generation then mastered is how to spread information to how to manifest. If our generation did do nothing for history, putting the word manifestation out more is spreading different ways to manifest has been spread at most by us Gen Z. He's interesting. The young ones, even the grazers we love, but we was not off. It was not off this come all the way from the 1800s. Nikola Tesla

said it best. is what actually happens when you meditate. Meditation is a conscious practice of observing reality objectively without thoughts, judgments or emotions. This is done by controlling your conscious awareness and your conscious attention.

Okay, losing weight can be hard so here's how I lost 30 I think it's us

broke matrix Angel numbers and coordinates they start to draw your attention people

call the ISP and the Sunday let me tell you about the Sunday but some people lived under the sun

still many more to go through in this live here we're going into the sea creatures and yes this is a mermaid Alex I might be on later tonight like like five six hours from now.

We'll see you know single dad busy schedule I try to get on live when I can.

Burger to good burger to Charlie has a quick scene in the movie

45 degrees in the comments. I'm not even paying attention you guys you guys are just triggered by reference.

This is how you figure out your purpose right now this message came out of nowhere. So make sure you listen to this and save it because this is 1,000% going to change your entire life. Want you to ask yourself one crucial question. One very, very, very important question.

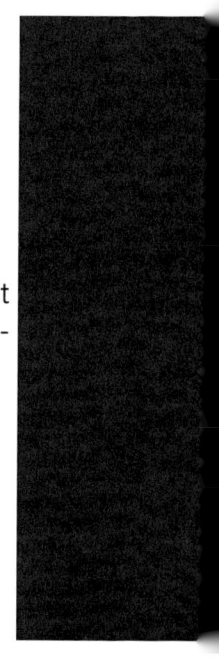

What if you were living your purpose right now? What if your purpose wasn't a destination, but it was a journey?

And what if instead of seeking to get to the top of the mountain, you realize the climbing it was the whole point of why you started in the first place.

Government is banning these supplements by 2026 for their potential to cause mass destruction. Number one, magnesium glycinate, known for its ability to put you to sleep faster than melatonin and produce the most insane and vivid lucid dreams. Number two moringa powder comes from the leaves of the moringa tree, also known as the miracle tree has 10 times more vitamin A than carrots 17 times more calcium than milk 15 times more potassium than bananas, and 25 times more iron than spinach. Number three NZT 48 This is the most powerful brain booster to ever exist. Just a single pill is 13 times stronger than regular Adderall. And we'll unlock 100% of your brain power for 24 hours once taken.

I got a message from your grandmother. Okay. My name is Marguerite, Mildred. Amy and Annie. My lifetime is Carter. Okay, you're going to see somebody in a costume. I know that is not Halloween. But this person has a performance or something. They have a performance that they're going to have some kind of show, they could be on a train, they could be getting out of an Uber or something like whatever they have their own. Okay, this will be a sign for you. You're also going to find a bunch of change you're going to be see seeing feathers, you may see an ele-

phant somewhere like a picture of a significant elephant. Your grandmother wants you
to know that she's handling your Mom, don't
worry about that. Do not do anything. Okay.
You're about to be blessed with security, the
king of pentacles. You have the six of wands
here you came public recognition, the 10 of
cups, okay, she's wanting you to put your
guard now you've been doing something and
you're supposed to be doing something else.
She's saying that you're procrastinating. You
need to pray to remove the spirit of procras-
tination because there's a project that you
need to go ahead and get out. Okay, ace of
wands, this could be some kind of invention,
some kind of business that you're starting up,
but six of pentacles with the queen of wands,
or the king of wands here. You are about to
receive what you have been giving. There's
something that you recently did, I don't know
if you would help the homeless person or
something, family member or friend, you did
something from the kindness of your heart.
This person was not genuine and reception is
what I heard. They did not. This person really
got upset with you because they're mad that
you were able to help them. Which is kind of
strange here. With the eight of wands, you're
becoming the queen of Pentacles. The king
of pentacles. Very fast here. Queen of Wands.

She's saying that you're also about to get a burst of energy here. My name is Hannah or Haley or her Lena. Okay, I keep hearing Mildred Mildred. My name is Ronald Rodney or something like that. Janice Janet. My last name is Cooper. She wants you to know that there's an ending to something you've been feeling some kind of way. You've been feeling like you're down and out or like something is not going to work but it is. You're so strong. You have somebody that has been watching you. You're in harmony, this could be a verbal or you could be somebody has been gathering information about you, but I feel like they want to work with you. You're about to be getting so many offers here. Okay. It is so many offers that's coming for you. You're seeing a lot of crows. Okay, high priestess in reverse. You're not listening to your intuition. She also wants you to know that there is a low vibrational High Priestess in your energy. They have been doing something but she's seen me. She's seen it. And she's coming in to help you in some kind of way. Confirmation. Yeah, the three of pentacles somebody has been working together against you, but you're getting justice, whatever you feel like it's going on. It's exactly what's going on. But she's saying you need to get more rest. You need to be paying attention to your surroundings. Well, you're safe. If you've been worried about your children, they're safe. Your children love you very much or your children are very gifted as well. Somebody was going into spirit toss on block you but your grandma's right here. She's saying no somebody named Betty. Penny Penelope, Betty. Carol Carolyn Sharon. Okay. Camille was here. My name is Julie. Julie or Lily Lily's could be significant Daisy. What else? DayZ the critical servers, there's an offer that you turned down, you did the right thing, okay? Somebody was going to take advantage of you. Somebody is plotting all your life every single day, and your mom wants you to know that she's not playing when nobody, she could be a Leo here. I don't know what this is, but you definitely have a whole lot of blessings coming towards you. You're gonna see several different cars that are riding on doughnuts or something like you're going to be seeing a lot of cars they got spare tires on. For some reason, I don't know. It seemed like somebody was trying to tamper with your car. You may have a spare tire or something with your tires. You may have recently had to put air in your tire and I don't know where you could have.

They did physically exist. Yes, they physically

these were physical creatures. I do believe in magic. Yes. I do believe some of these things do still exist today. Especially given the one the ones out in the ocean because the oceans are so massive.

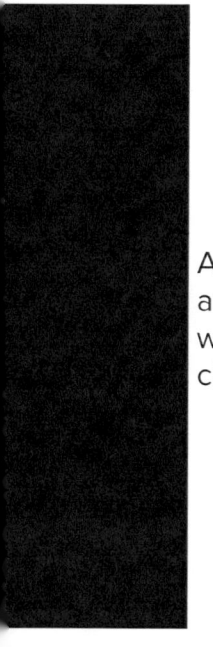

Alright, we got to talk about Sam and Kobe and Cody and Satori. If you have no idea what I'm talking about, take a look at this clip.

Already. A, B, C, D, E, F, G, H, I, J, K, L, M, L, A, B, C, D, E, F, G, H, I A, B, L i b.

Is this make sense for you?

Do you need a break?

No way to fucking have they know anything? No. That's not her real name. That's her nickname. No one in my family. No one besides my family. Anybody like that? What the fuck?

Since even Kobe or two YouTubers who are known for the paranormal content, they're massively viral. They have a young impressionable audience that trusts pretty much everything they say Cody and Satori are two psychic mediums. Satori is the daughter of Jason Hawes. Cody is someone who claims he has this like ability to speak with spirits as you saw Sam and Colby were absolutely blown away by the knocking and the seance they held this video is specifically for you saving Kobe and their followers. You're being fooled before you say Oh, who are you to judge?

Let me go over who I am and my qualifications. My name is Dustin Dino professional mentalist and psychic debunker have appeared on national television debunking psychic powers on Penn and Teller fool last night, we spent the last 10 years traveling across the US performing mind reading illusion shows and even fake seances. But I'm always very open about everything being fake. And just for entertainment. The point is, I'm an expert on fooling people, which means I also know when someone is being fooled. Now I appreciate Sam and Colby, taking some effort to try to debunk what they see. But the problem is that they aren't experienced in the techniques that psychics would use. Then How the heck are they going to debunk it?

And I know this because the way they debunked it was a very poor debunking one point in their debunking video, they asked Cody to take off his shoes. And he does he's wearing socks, but he won't go any further than that. He won't do it barefoot. He just laughs it off and doesn't a lot of the things they requested he just wouldn't do because he said it doesn't work that way. In fact, one of the things they said is they have to hold hands coding Satori, to make this work. To find out later that about 11 years ago, Cody was seen in a video doing the same exact method of debunking couple of the major things we saw on this clip right now very easily. The first we'll get out of the way is the grandmother, how did they know the grandmother's name, how they get the nickname well took me about two seconds to find Sam's grandmother's obituary just by typing in Kansas, which is where St. Louis Brown go back, which is the last name and that obituary, and that immediately popped up with all the information about not only the grandmother and her nickname, but all of his family members and things about them. Cody and Satori obviously knew Sam and his last name and a lot of information about him before they started.

The second thing is the knocking. Now, a lot of people have debunked this already and talked about the fox sisters, Sam and COVID even mentioned it themselves. boxers were known for faking seances in the early 1900s. And they were known for causing what's called rapping. That's not the singing rap. That's the knocking sound. You hear in seances or they were known for causing this noise using their toes or by cracking their joints. So a lot of people have been demonstrating that I do not believe that is what Cody did. That is a very old technique and it would fly back then but it wouldn't really fly. Now. In fact, Sam Hobi explained that they could even feel the vibrations but I'm going to show you exactly what I think he used and it's something called a solenoid.

solenoid is a device that you can plant anywhere. You can hide this on yourself. You can hide this under the floorboards you can hide this in the walls, you trigger this with a toe switch, and it makes it knocking noise. Let me show you Kevin. Leave that here. I'm gonna stand way back here. Hopefully you guys can still hear me. No one else in the room. Maybe it's just me and my dogs. I'm gonna leave my hands up here. We're going to have my feet everything so you can see I'm not doing anything. If there's a spirit here, please give me one knock for yes, two knocks Are No? Okay, well, that's a little bit confusing. That was two knots. I'm not sure what to think. Okay, here's how we did it. In my SOC, connected to a power source, excuse my feet, we have a toe switch. This toe switch activates that knocking noise which I clipped in this wooden plank. To simulate a floorboard, our source is just a small little battery pack. And the thing that makes the knocking noise, again can be hidden anywhere can be hidden in your body. It can be hidden under the floorboards on the wall, anything, a lot of people were saying like it can't be debunked. There's no way. There's a million ways, this is just one of them. But here's the thing to erase any doubt in your mind if they can do this method under my conditions, and if I can't explain it, I will give them $10,000. And I'll admit, this is legit. It would be undeniable proof. I'm not too far from the conjuring house. I'll just take a drive. And I'll meet you there. And you can just do the same exact thing that you are doing for everyone else that you get first name and Kobe multiple times. If you do it again, $10,000 Easy money, right? Can't wait to see the response.

There's only nine numbers in the universe. Nine numbers. 10 is actually one plus zero. That's what zeros are placeholders. That's a man made creation. Creators language is math. It's only nine numbers in the universe, one through nine. The numbers have a resonance, a frequency a deep sacred geometry. And that this three sixes and nines is special. Very special. Who said that? How about last century's greatest genius. Not Einstein, Nikola Tesla. Tesla taught his students John Kelly of oil, Raman rife and others. If humanity only knew the powers of the three sixes and nines, it will be a completely different universe. That just world universe said tested. The Alpha numerics alphabet numbers now the Numerix of the English language, one through 26. But we talked about earlier, there's only nine numbers. So you got to use the factory and mystery school math method of reducing the double digits into the single digit integer. So you take one through 910 is one plus zero is 111. One plus one is 213-913-9128. Then he took the words faith trust

and God. Oh, wow. Isn't that interesting? He noted they're all eight, the infinity sign Gods number nine is Completion. Now you might say Oh, well, that's nothing fancy. There's nothing just interesting, but it doesn't prove anything. Here you go.

Rockefeller Rothschild cartel, that they that they deceived the wealthiest men of all the nations of practice sorcery, the word of sorcery is Pharmacopeia Pharmaceuticals is the root word of sorcery, and the cast magic spells to do it, it says that's exactly what the hypnosis machines called the television are casting magic spells indoctrinating people into the cult, to the deceptions.

Thank you to the Rockefeller cartel is operating on behalf of the royalty of Europe and the banksters of Europe that are running Carnegie and Rockefeller, institutes and foundations. The fundamental mathematics of this starts with the whole note, the whole note and music reflecting the whole cosmic system. Now imagine you were God for your imagination. And you were all that was boring. You'd be bored. Moreover, think about this for a second. If you were all it was, if you expanded yourself as let's say you were God in all directions to infinity in all directions, massive outward growth and beyond beyond until there was nothing else but You, there would be nothing else. But you that means everything and nothing is the same. It's all it's all that one symbol. That's what that symbol represents. Now the shortest distance between two points, the creator and you has a straight line. So you divide that now you've got the rudimentary eight, which is the infinity sign on its side. You see, 45 degrees, 90 degrees, 180 degrees, it's all Masonic.

It's all architecture. It's all sacred geometry.
It's all very mathematics.

I'll watch it. Like every notification I'm getting is you guys tagging me this video,

we let's we gotta check it out. Call me crazy,
if you want. But I've never liked

Starbucks, Susie. All right, that's absolutely enhanced. So this one night, I was 21, I was going to sleep like normal living in Wellington, New Zealand. And I was expecting to just get up and go to uni the next day. But I ended up falling asleep and finding myself in a dream. Now what typically happens with me when I'm dreaming, as I'm aware that I'm dreaming, I don't think it's quite lucid dreaming, but it's just like a level of awareness, I guess. And what was starting to happen around this time as well was I was getting pregnancy dreams. And this happened to be one of those. So I was pregnant, and I could feel the baby in my belly, I could feel everything. And you know, got attached to it, because I'm a young woman, and it's biology. But after a while, I'm starting to realize that the stream is going on for quite a long time. And it's getting later in the day later in the day, to the point where it's time to kind of go to bed, and I'm deciding this is weird. I've never gone to bed in a dream before. So I go to bed, expecting that when I wake up in the morning, I've just been woken up from a dream. I'll be back on my bed and Wellington.

But that doesn't happen. I wake up, and I'm still in the train. So the cycle kind of repeats itself over the next few days, I kind of wait to go to sleep at night to hope that I'm going to wake up again. And I don't wake up. I'm still just in the stream. And I'm kind of thinking about my family. At this point. I'm missing my boyfriend, I'm missing everything. And the more days go by and more days go by and I'm in labor. And I guess it's good to say that like an all their pregnancy dreams I've had before. I've never given birth in any of them. So this was kind of like, the moment I was waiting for. When all this was going on.

I was like okay, well, I've never actually given birth in a dream before. So when I'm getting birth, that's not my cup that today. No, I was on the hospital bed, I started giving birth. I've also never felt pain in a dream that I felt this I felt so free that of the labor. And I gave birth to a baby boy, I named him I did everything that a new mother would do. Which is weird, because how would I know what a mother would do? I'm not a mom myself yet. And every day, I was still kind of hoping to wake up still, like the kindness in this weird little space. Like starting a new life, but also wondering when I'm gonna get back to my old one.

It gets to my son's second birthday, and I reach acceptance. I might okay, I miss my family. But I have to accept I'm never gonna see them again. This is my new life. Now I have my son. This is my life. I watch him grow up. I watch him go to school for the first time I watch him get his first crush, graduate, get married, have his own two kids. And then on the day of his 40th birthday, I throw him a surprise party. It's kind of weird to say there was never a father in the picture. Like it was always just us too. But I remember what it looked like. Like there was a picnic table outside. There were heaps of trees in our backyard. We lived in a double storey, wooden terraced house, and we're singing him happy birthday. And I woke up. I was back in my bed and Wellington. No longer this like six year old woman with a 40 year old son. I was 21 and my dad about to go to uni. Like I was wanting to for what felt like 40 fucking years into a dream that as I laid out when I woke up, I was happy to be back. But I missed my son. And I cried and I went through this grieving process for my son that I had raised for 40 years. I couldn't tell anyone because I don't want them to think I was crazy

I don't first of all, it gave me like Inception vibes. Second of all like what A fog that is insane to me. Insane. She lived like another 40 years, and then came back here. Like, I literally don't even know what to say. Like first

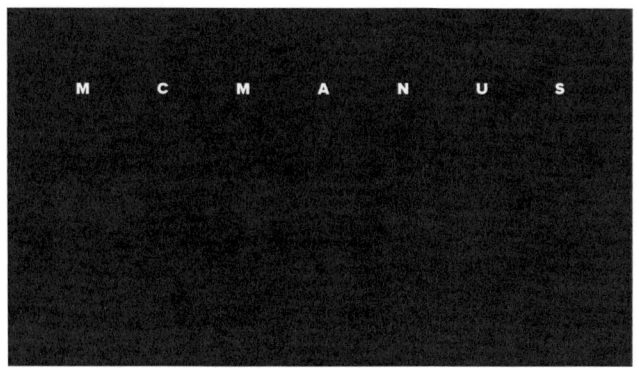

initially, it was like, Okay, maybe she died in this life. So she then she hopped into another one like a quantum immortality thing. But then she came back 40 years later. So she basically just like, hopped into like a parallel universe reality situation, but stay there for 40 years, and had a whole fucking life. I can't even imagine, I can't even imagine waking up in that reality and dealing with that. And then 40 years later, coming back to this one, and then grieving all of that, like,

what? Liquid? I feel like I'm saying nothing.
I'm literally just saying nothing, basically.

And how do you

not tell anybody?

Like how do you come back here?

And you tell anybody, and you have that in here? For so long?

This would be really fucking stressful. I want to give this girl so many hugs. And they're like, when she was in this other one. They're like, Did she ever say to anyone in the other reality like, hey, this happened to me, and I woke up here, and I don't know what's going on. I'm not supposed to be here, or does she keep it inside there too, like Did she literally has, she kept all of this inside for 40 years, plus, whatever else when she came back until now, when she just shared it with the world's friends. Holy shit, it's no wonder you guys could tag me in this thing, okay.

In order for you to find that if somebody hates you, or somebody being fake to you, what you do is you're going to get a pot, you're gonna get salt, and you're gonna get a Black Candle, make sure that you always have a candle lit, every time you do a spell, you're gonna put a salt on the pot, and you're gonna draw the size of the power. You see, you're gonna save the words that I actually wrote on here. And you're gonna hear from that person within 24 hours asking you for a favor. That's how you're going to know who those people are. Hot

take not. This is why you should never listen to a channeled message or a reading for the collective, nobody can connect the energy of your life better than you can connect to it. I don't care if you have no experience or where you are in your spiritual journey. No one can do it as good as you. So why would you trust anyone else with a message, especially when they say things like, take what resonates and leave the rest. When you're tapping in and reading for yourself, it all should resonate.

I feel as

though this chord with that is supposed to be attached between the physical body and the ethereal body,

not the physical by it's called the silver cord. It's a pulsating, excuse me a silver cord that's attached from the back of the astral of a spiritual body to the higher self. That's why it's pulsating, you're getting energy from your higher self. Now in a regular out of body experience, most people don't see it because it's behind them.

Or I have to kind of spin around to do that.

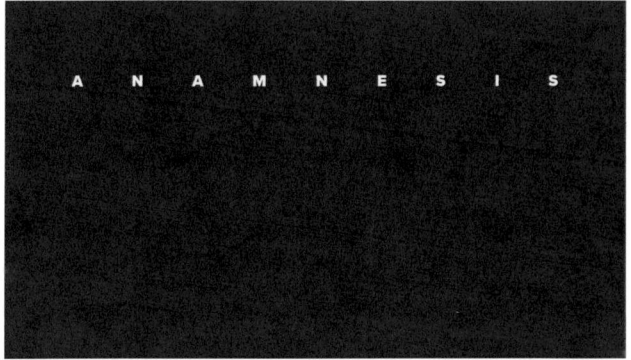

We'll see somebody else stretch indefinitely, right?

Yes, but here's the difference when a person clinically crosses into spirit or clinically dies, that cord is severed, severed. So if you see a similar cord and in borough exactly except you to death rather than birth. So if you severed you ain't coming back, because what the point is on that, but it's pretty bright eyed about the experiences are not dangerous. They're perfectly natural to be severed by mistake.

No, there's a reason for it.

You It's not like somebody has a sword swishing around, they're doing that clip you there's a you choose your moment of death and the death is actually well planned. So it's not you're not going to accidentally go out of the body and sever the silver cord that has never been recorded. And I've done over 3000 times, and I've seen my own silver

cord to what have you seen when you've done your own? Well, I

go into the fifth dimension, I go to various dimensions I've described. We spend a few

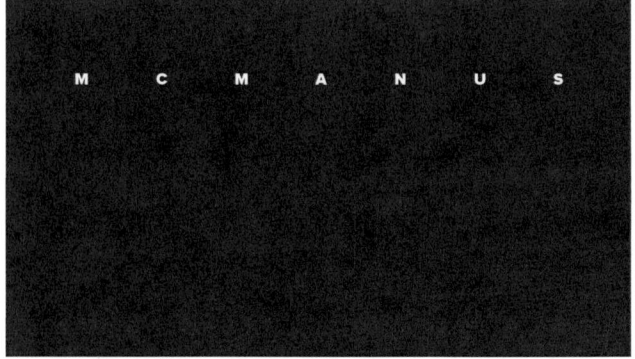

seconds talking about that.

You have the astral plane, every dimension has a different sound. The astral plane is characterized by the roar of the sea like a seashell to your ear. And if you go to the astral plane to the upper not the lower one, the lower one is psychic attack that you offer would be godlike. It's not godlike. It's just very positive.

Okay, you have the astral Museum. That's where Bill Gates and Einstein went and Tesla and everybody else to get futuristic technology. They brought it back. And that's how I developed my superconscious mind tap in my progression take things I went to the National Museum to the lower astral plane is filled with negative entities you don't want to be there demonic entities somebody to be called. Yeah, they're called more purgatory be the better analogy because you can get out of it. Because that permanent okay.

Now the plane beyond that is called the causal plane, which is sound, the sound is the tinkling of bills like a new age bookstore. And the causal plane is where it's a very small place the smallest dimension, and that's where your Akashic records are stored. Beyond that is the mental plane, which is the

sound of running water characterizes that. And the mental plane has blue highways. So if you're driving in a car or flying above it, and you see iridescent blue roads, you're on the mental plane, all the entities or the people that are there, they look like little angels.

They live in these houses of geometric designs, and the Masters would hang out there the Buddha's the Jesus, the Moses, they would hang up the mental plane. It's a very spiritual dimension. Beyond that is the etheric plane.

That's a biggie. The etheric plane is the largest dimension the sound of buzzing bees is the characteristic and that's when I talk with you on the show about the theory plane surgeons.

So if you have a medical issue that like cancer aids or you lupus or something that you can't cure, and you are there's nothing you can do on the earth plane.

The reason why the etheric plane surgeons work well these are entities that work on the etheric plane. Your authentic body is identical to the physical body. All the other spiritual bodies are similar but not exactly identical.

The etheric body is identical. So, if these authentic searches look like giant tadpoles, if they work on your etheric body and cure the etheric body, then not a high percentage, but in certain cases, you bring it back to the earth plane, when you awaken the following morning, you can transfer the healing energy and I've done that with patients.

So this is called etheric plane surgery. It's a very ancient technique. It's been around for many years, the old mystery schools in ancient Greece and Rome and Egypt. But I've actually westernize it for people to use. It's not a high percentage, but you got nothing to lose. And when it works, it's brilliant.

Before I get into this video, if you're sensitive, keep scrolling. Most of my followers know me from when I used to be a Hebrew Israelite. I am finally making a video in depth why I left? First of all, y'all are so big on using white man's religion, white man school systems, but you're using a white man's book, The King James Version, one thing I heard,

I

would like to know what happens to a to a fetus a child, when it is aborted

first of all, understand that the Spirit the soul cannot really fully connect to the body for 49 days. That is the time at which the pineal gland forms and once the pineal gland forms, then the idea of the expression of consciousness can come through the body you understand. Nevertheless, understand that the spirit that being that was representative of your child knew what your decision would be and only wanted to stick its toe in the water of physical reality.

Okay,

you understand?

I think nothing happens by accident. It's an orchestration that beings still exists still around you still loves you. Still perfectly viable? knew that it didn't necessarily want a full physical life knew that it might just want to give you an opportunity to go through this process. You say no? Okay, though, do

you listen to real God is love God and Jesus is?

Okay. So Jesus is real. That's another question a lot of people wanted to know about.

Deepak do anything for you write a story about life after apocalypse. This is Shadow one. Once you download this app, you basically get a personal AI assistant, say the words in this thing can create your wildest dreams, create essays, speeches, and more channeling delivers human like interaction in real time, this app is blessing, you can get it at the link below.

Oh, this is going to be controversial. So many of the religions are worshipping the wrong quote chap, the bloodthirsty arrogant tyrant of a God described.

So if you're following my channel, you would

know that words, cast spells. And the one word I want us to cast the lay is the word release nothing about the vibration of the word. When you read anything, it usually means that you are going to do that thing again. So think about it. If you redo something, what do you learn to do? You're going to do that thing again. So when you re lease something, what does that mean? You're going to lease it again. And that's why they say the English language is a spell. Because whatever you're trying to cast away from you, you're actually calling it back to you because you're leasing it again.

If particularly tired then

really try to come up with chromosome one. So the scans done mode. Okay, Chang, I have no pink car loan officers by Carlos. And there's no real reason the real world

really tried to come up with a clever way to start this video, but fuck it. I don't even feel like let's bring the number nine to the lightning in the rain pace ever changing your

dialogue. This book actually goes into narratives and creating new ones. So

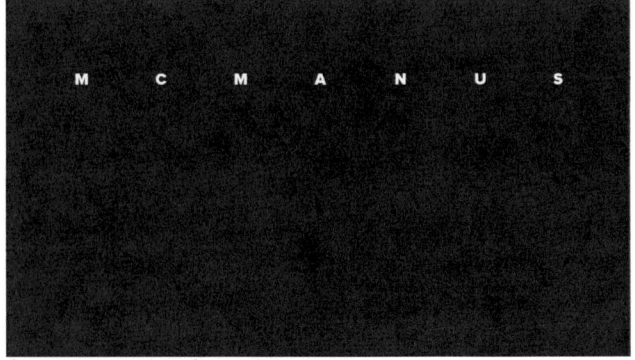

for example, five things

to never do in the spirit world. Number one, play around with the Ouija board. You don't know what portal you're opening. You don't know what games are gonna come through. So just don't do it. Number two. Never have a welcome mat. A welcome sign outside of your residence. You're inviting things that you don't know flee exist into your space, into your home, into your family's home. Next, never sleep with a mirror facing your bed, never face a mirror towards your bed and go to the you're defenseless. Never have two mirrors facing one another. You don't know what will enter. If you don't know what will enter, then that means anything can enter. Next, if you hear a baby screaming, something is knocking at your door or on a wall, do not go to investigate to see what it is. If you're not trained, don't do it. Because oftentimes, this is baiting you into a trap spiritually. And finally, if you want to challenge a spirit, if you see something or hear something, or feel something in spirit, and you want to challenge it, and you're not trained to do it, you need to stop and pray.

There are no coincidences. In our tradition, what we in the traditional language speak about is it's called karma. Many people maybe heard this word already. But karma just means if you want one day to have orange tree, somewhere in the past, you must plant the seed for that orange tree to grow. If you want apple tree, you will need to plant the apple seed. In this world, it just doesn't happen. That we are planting apple seed and get orange tree.

So the first observation is first of all, to just look at yourself sometimes and and take this status quo. How do I feel right now?

Where do you see like your areas where you could still improve yourself?

And then try to relate this to everything that happened in the past, to your behavior in the past, to your thinking patterns in the past?

And just based on this type of observing yourself, taking a moment and watch yourself, what are you talking?

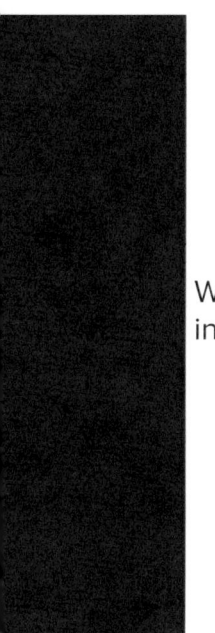

What are you thinking? What are you saying? How are you behaving just based on that.

Normally, if you watch carefully enough, you will already start to build up a connection between the past and who you are right now. So you start to actually realize that you are exactly where you have to be right now. It couldn't have been different.

What I'm about to teach you is going to be the most important thing that you've learned on tick tock today, tomorrow, yesterday, regardless, especially if you are someone who is trying to change your life and manifest great things. If you feel like you are experiencing so much stagnation and resistance in your journey in your path forward in your life. Listen to this.

While so many people talk about the law of attraction on Tiktok no one is actually teaching you how to implement it. And when I always tell people is you have to do the work first you have to do the work on yourself first and then you'll see it happen around you. And this is actually known as the universal law of correspondence. I will live and die on this hill about this universal law because it is more important than the law of attraction. The reason why it is more important is because you have a set of beliefs internally subconsciously that dictate and control your life they control the way you perceive people perceive yourself how you are allowing yourself to grow and change and react to your environment. You might have heard the famous Same as above so below as within so without and it cannot be more true this law essentially states that the internal state that you have within reflects what you see without

Hey guys, welcome back. We're jumping right in. I'm sure after the last video you have a lot of questions. I'm gonna continue to explain the New Earth jumping back into convoluted universe book to the New Earth is another dimension, we will move into a new dimension. In the last part we talked about if you are on earth at this time, which clearly you are, if you're watching this video, that your soul chose to be here at this time to be a part of this shift this evolution of the planet into a new dimension into a higher consciousness. Those who are working for the earth for the universe will be provided for and will continue to be, what you need will come to you. So it's time to let go of the ethic of working to get money, you are working to change the earth, your work must come from love and service, not from greed and convoluted universe, they talk about the two types of ways that the Earth can shift, the crust could shift and then you would start from ground zero. That's what killed off the ice age and killed the dinosaurs, then there's times that there's been dimensional shifts, they say that Atlantis and Lemuria experienced dimensional shift. There are groups that have done this before including the Mayans why they just happen to disappear, and other civilizations that are surrounded with mystery that simply disappeared leaving no clues of what happened to their civilization, they raised themselves to a certain frequency, and their consciousness expanded and we were not able to perceive them in our reality anymore. Those who had raised their frequency and vibration would ascend into a new earth as it evolved and lifted into a new dimension, thus becoming invisible to those left behind it goes into the Theory of Parallel Universes, Dolores gives a very simple explanation. Anytime an individual has to make a decision, this is called coming to a crossroad, they will decide to go one way or another whether to get married, whether to get divorced, whether to start a job whether to leave their job, when you put thought into making a decision, you give energy to both realities, when you make the choice to go down the path of one reality that becomes the world that you perceive. However, all the energy that you put towards the other decision, that energy is not going to be destroyed, because you didn't walk that path. So that energy continues to become a reality. It says this is a simple explanation. But doesn't only happen with major decision, another version of you always splits off and plays a different part. Our human minds will never be able to handle it all. And it says the way that we're not able to perceive the other versions of ourselves, say you quit your job. There's another version of you that's still working at that job and didn't quit. You're living your life without that job. But another aspect of yourself still exists at that job that you wanted to quit. But you're never gonna be able to perceive the reality that you didn't quit that job. And that's what would happen with the New Earth on a grander scale. The people on the two Earths would be unaware of each other. And we're going to get into this all of the chaos and turmoil that is coming. But here it says that the people who came to Earth at this time need to stay grounded and tranquil for other. This is why a lot of the older souls who are here to work on this earth shift, experienced childhood trauma and abuse so that they would be able to be a pillar for others during a time of chaos. Those who are here now need to remember the important role they are playing.

Hey

guys.

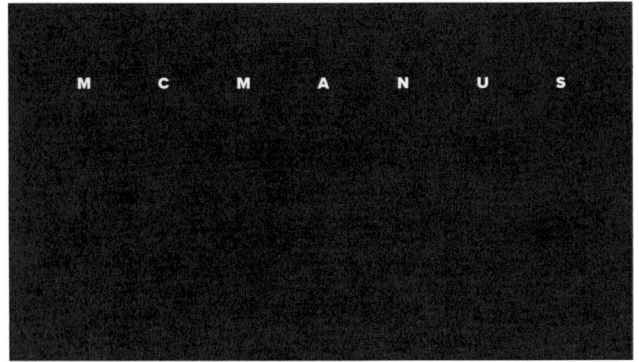

Be careful. You have a snake around you. This snake comes around you and they hug you and they greet you. And they talk to you. Like they care about you so much. Behind the scenes with third parties a plot against you. At

around two in the morning, a man named Daniel Palomino was walking alone inside a plaza when something completely unexpected happened. It was actually during the live stream when Daniel recorded this. So this is something that he often does during his free time. During late at night. Daniel walks alone all simultaneously live streaming on his phone. However, what happened on the night of September 28 of 2023 has scared him to his core. Like any other night, Daniel was live streaming to an audience who were suggesting places he should visit during his walk. One place that was recommended was an old Plaza that's believed to be haunted. Although hesitant at first, Daniel heads over to the plaza after realizing how close he was to the place. But this was a big mistake. After exploring the strip mall for more than a few minutes things started getting weird when it came out

M C M A N U S

when it came out in a fishy

strangely, Daniel begins to smell a burning scent coming from somewhere in the plaza. He tries to find the source of the smell but gives up before suddenly the scent goes away. Although pretty strange, this occurrence becomes a whole lot creepy after one of his viewers on his Live Stream tells him this

no matter the discipline, the central agenda caning Aquarena tallinna racy and I guess you know everything you own in Sandy Maria don't personas not not knotting.

According to the viewer there used to be an old casino inside the plaza

they caught fire and took many people's lives. Hearing about this left Daniel totally freaked. He just smelled the burning sin and thinks this might have been somehow connected to the fire incident that occurred years ago. Feeling spooked. Daniel immediately starts to head out but on his way out something in the distance makes them stop in his tracks. This is what he saw

his

issue?

Key.

We won't get to

a mythical mythical. We

prepare to see a lot of people shapeshifting Yes, you heard that right. shapeshifting. You know how there was a mass spiritual awakening in 2019 and 2020. That happened to me as well. I was atheist up until 2019. Then all of these people, including me started to get serious spiritual downloads, we started learning about the secrecy of the universe, our galactic history. And we also started developing psychic gifts.

I feel like this is happening every three years. It's a three year cycle. And we are approaching the end of this year. And I feel like there's a whole influx of new people who will start to get their spiritual gifts. And one of the most powerful spiritual gifts you can see is the spiritual gift of sight. But it's not your two eyes, it's your third eye.

And you know all this talk about the Age of Aquarius, while we are approaching the Age of Aquarius and I have come to realize or theorize that every time there is a change in the ages. So we're going from Pisces to Aquarius, there's almost an interface of portals opening. And with these portals opening and the sensitivities of people's psychic abilities starting to rise, not only are these gates opening, going to give you the downloads to your spiritual gifts, but it's also going to allow the veil of reality to kind of thin so the matrix reality to thin out a little bit. And a lot of you guys are going to see the real truth of this reality, which is that there are different kinds of alien species living amongst us as we speak. And they have been here for hundreds of 1000s if not millions of years, they've never left. If you want to learn more about our galactic human history, I've written a book and that book is linked to in my bio truth is stranger than fiction. But my prediction is be aware of mass unrest. And I'm not just talking about protesting or rioting, because the government is a piece of crap.

But I'm talking about people feeling like their mind is going crazy, like they're suffering from schizophrenia or psychosis, when in reality, it's just their spiritual gifts starting to fill up. And I know that if you're spiritual, and you've gone through the same thing, you also thought at one point in your journey that you were going crazy. As long as you were functional, and you're not hurting anybody, you're not crazy, you're just experiencing different levels of psychic gifts and intuition. This is also why the US government has been slowly kind of leaking out little clues about UFOs and aliens and how they've always been here, it's going to create a domino effect in which a lot more countries are going to start talking about the same thing. Now what they don't want you to know is that humans are also an alien. We're all hybrids of different aliens. And when you get the spiritual gifts that are going to come in, I think by 2025, at least a good 25% of humans will start to have this gift. You're going to freak out a little bit as to who somebody's higher self or true galactic lineage really is.

I was diagnosed with alopecia areata at the age of eight, and I've had my hair grow back and fall out three separate times. I've tried many treatments over the years and although I haven't tried a loom yet, I'm excited to share that it's a treatment option available for adults with severe alopecia area will

hug a tree if you want to feel good, that's not nonsense. A tree roots go deep into the ground. They're anchored. They're spread out the branches, the roots and the energy that comes up through the tree. When you hug a tree. You're you become part of that energy. And people say well, I didn't feel a thing. Well, yeah, I understand that. But look where you're coming from. How many trees have you hugged? And would you hug a

tree? If somebody wants you hugging a tree? Would you feel comfortable enough to to still do it? You see, there's things you have to do to overcome, go hug.

If you don't believe in Angel signs or signs from above or anything like that, keep scrolling, because you're gonna think I sound really stupid. And I'm not even convinced that I don't sound stupid. So I only want somebody to tell me what it means not tell me it's not real. So, basically, a year ago, I switched phone providers and I got a new cell phone number, I went into bail, and I said, Listen, I'm with Rogers hate them. I'm going to need a new phone and a new phone number. So they put a whole bunch of phone numbers on the screen. And they told me to pick one, I picked the one in the middle because it's the easiest to remember. And it has the most common uncomfortable, has the most common numbers for me to remember. So I pick the middle phone number, and nobody knows I have it. I didn't tell anybody went to the phone store. So I get all set up with my new phone. And I go back to my car. And like I said, nobody knows I went to the phone store. So I'm like, I'm going to text my boyfriend and just mess with them a little bit. So I say like, Hey, sexy,

and he's like, Who is this? And I'm like, huh, like guess she's beautiful. And she like needed a new phone and all this stuff. And he's like, Jenna, like, this isn't funny. If this is you, like call me.

So I call them and he's like, FaceTime me. So FaceTime anemones, like freaky and like all freaked out and stuff. So he comes over, and he shows me his phone. Guys, are you ready for this? Like, are you kidding, huh? guess she's beautiful. Her phone broke. So she went to got a new phone today. Grandmas cell. I literally picked his grandma's phone number who passed away the year before? Like, I've never believe believed I can not speak today. I've never believed and like any sort of sign from above or anything, but like, that's really freaky guys. Like, I had to tell everyone in his family about that. So when they get a text for me, it doesn't come through as like grandma Ruth or like mom or aunt or something like that. I don't know. Like, we kind of think it might be a sign of her being like, what up? So have you heard about this supernatural phenomenon that sometimes happens to people in really bad car wrecks?

I was in a bad car accident. And I've always had this strong intuition. I've always disliked it. But this day I did not. I'm driving. And usually when this intuition takes over me, it's about somebody around me. But I was driving alone. And I'm driving and it's raining. And I'm like, oh, no, something bad is about to happen. I'm having that feeling. And I'm like, Oh, I'm by myself. Let me just pay attention to the road. And it got so heavy on me. Like that's so heavy. I'm like, oh, it's about to happen to me. And I started to think about what happened to me. I'm driving, what and the only thing that could happen. And by seeing a car accident, I sit and I just put both hands on the wheel. And as soon as I put the hands on the wheel, I always blast my music, right?

The music stopped. I started to pray. And I was like, God, please cover me in his car. And as soon as that sentence came out, I remember like it was yesterday. This was five years ago, as soon as I said that, I heard a thump. I didn't I didn't feel it. I heard it. And I said, Oh no, I've been here. I don't feel it, like what is going on. And at that moment, I kind of felt myself lifts in the car. So I was in the car, but I wasn't feeling any of the impact. I got hit by four cars. They pushed me down the interstate, like a hockey puck, like ping pong, whatever, like they were just pushing me down. And the whole time I'm holding the wheel and I'm like God, please. And I'm praying, come to a stop. And then the dust settles. And I try not to cry every time I tell the story. Hold on.

I'm gonna tell the long version to the story. I've told it before.

But the dust settles, I stop. And all I see is car lights. Because my car is it's like not parallel to the interstate. So all the cars can t boned me at this point. So everybody's in shock. I look around and I say my TED and my life like what is going on. And I'm holding on to a wheel very tight. And this is when I used to work really long nails. And I'm holding really tight to the wheel. And that here's somebody to the left of me. And I look to the left is some guy I can't tell you his nationality or race like he looked like a little bit of everything. But all I remember is he had on like a brown outfit. It looks like he worked for UPS. He had on a hat and then like curly hair. His hair was brown and curly. Nice eyes like they were brown. I just can't tell you what his race was. And he said he tapped the top of the car and he said hey, you're okay. Look like you're okay like almost It almost like, look, because let me you're okay. And I say I'm okay. Because he said he told me to repeat after him. He said,

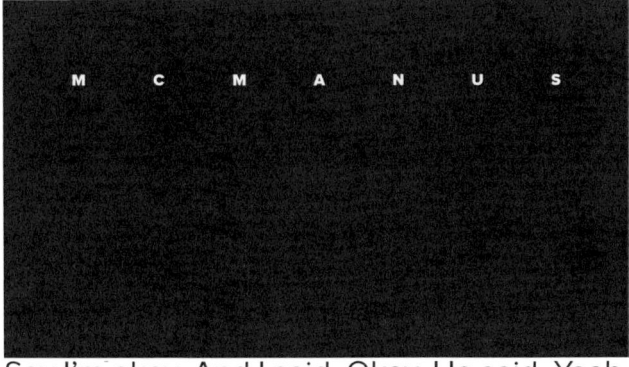

Say I'm okay. And I said, Okay. He said, Yeah, that's right. And like I started with a cry.

Yeah, I promise I blinked my eyes. And I opened them back and this man was gone. Like, gone. So the police come after a while, and they're asking me what happened. And they're like, yeah, it was a hit and run. And I really want to thank that man that came over like to calm me down like the first man that walked over, like, where is he?

And the cop like, what, man there's no man here. It was already like a patrol car already on the interstate. So that's how they got some nice affairs. But he's like, there's no man here. Nobody got out of their car. It's like I heard a little voice. I was still clinching the wheel when they came in my car was beat out really bad. Like, like, I couldn't get it. They had to get me out of the car. So something voice was like, you can let go the wheel now. And I'm just looking at my hands. My nails are broken has like, y'all. When I got to the what is it called the junkyard to get my belongings like after the accident. I walked past my car so many times because I didn't recognize it. And it's time for me to walk walk out of the accident with two broken nails. And like it was like scratches right here. It was unbeliev-

A N A M N E S I S

able. But I never like that to make sense of that experience with that man, like.

M C M A N U S

who is he? Where is he? Where did he go? I
didn't make that up.

I was telling you, I was sitting in my car. That man walked in it. I wasn't afraid of him. I wasn't scared. I knew he wasn't there to harm me. It was like I just accepted him and his kind words. And he just felt safe. And that was a scariest thing to me. And I was so overwhelmed with emotions. One because I knew the accident was coming and how everything just happened. But when I encountered that man, I kind of felt like like crazy. But I couldn't make sense of it. So when I saw this video, I was like, oh my goodness, wow, that's really a thing. And so that's my story, I'm gonna make a longer story. Because that entire day kind of like told me what was about to happen. Like, my entire day was lined out trying to tell me that I was going to be in a car accident later, but I just ignored the science. But that's my story about the Thera band syndrome. I think the new weapon is so happy five years with me and my current card. I love him. Thank you, God for giving me this bar. And I start crying again. But that's my story. Dating on

the app is iron. But when we talk about iron, we have to be very careful. Okay,

I just want to warn you before you watch any further if you want to continue taking the blue pill or the red.

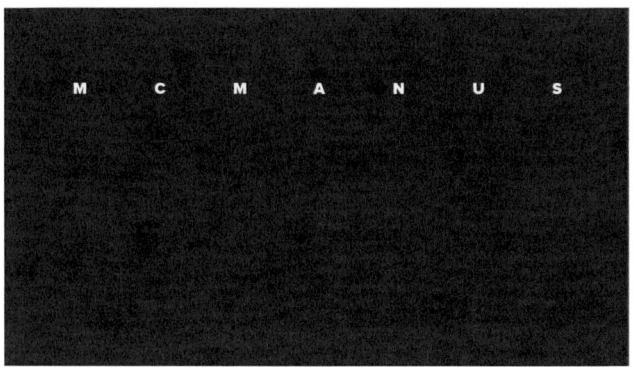

What's the what's the happy pill? Just I wish I could go back a month before I discovered tick tock. Oh, look at this. Cue the remote control turkey. Right? Totally. Oh my gosh. Please. Someone in the bushes is operating. I think look at that. Look at that. I know what's going on. We know what's running from us because they know anyway, I'm at this thing caught up our course. Yeah. I've ever been to suburbia. They make things like this. This track right? goes in a circle. And then there's like little spots to exercise on it. Things get crazy out here. And here's the thing. This is the back there

that is to the mom with a baby in the stroller

you know what it is? Every day, same time we're about to walk across the bridge. I've been coming here every day for a whole week. It's exactly the same thing.

Some of your

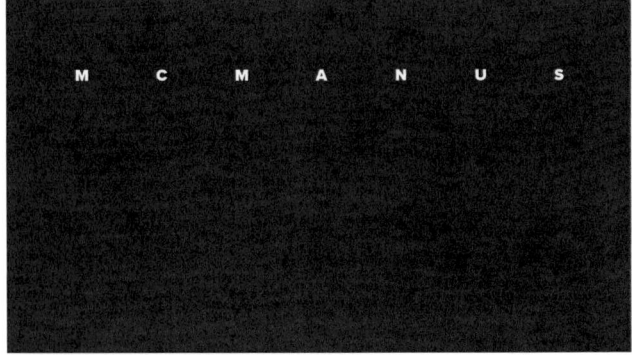

footage submitted by Jason something happens that if it weren't caught on dash cam you wouldn't believe it. It was around one in the morning, Jason was on his way to what he thought was going to be the house of a woman he matched with on a dating app. The person had given Jason their address over the app, they had not exchanged numbers. After putting the address into his phone, he realized that not only was the address 20 minutes away, he was also in a less desirable side of town. Against his better judgment he got in his car anyway to head to the given address, which was in a cul de sac. After entering the cul de sac, he realized all three houses had all their lights out. And when he tried to message the girl he matched on the dating app, she would no longer answer. After waiting a few minutes for an answer, he realized the person had unmatched him, meaning it may have been a trick or prank. And so he turned the car around. As he was leaving the coldest sack, he noticed someone on the sidewalk stepping out into the street. And so he locks his doors and slows down. This is the footage captured on the dashcam.

We talked about Anunnaki. But there's also hybrid Anunnaki humans living on Earth with us. There's also reptilian human aliens living

amongst us look like us. So maybe they look kind of weird. Hi, D matrix.

I'm gonna follow up yours for a couple of months now, and I really want to share my experience with you. I've always been in tune with the supernatural, but never anything psychic. Just being able to feel energy shift or see spirits that want to reveal themselves like in this story I'm about to share. It's a bit long, but I really want to hear opinions from people who don't think I'm absolutely crazy, because this definitely sounds crazy. And if I was anyone else, I wouldn't believe someone telling it to me. The only reason I know it's real is because I wasn't alone, and I had other people there to experience it with me. This sounds like it's gonna be a good story. Indicating be aggressive unless you're a demon. When I was about 13 or 14, I decided to visit my friend's house for the first time. We've been friends since childhood at drifted apart through fifth and sixth grade but reconnected in the seventh grade. We'll call him

Come on, take it,

steamers steam

it smells like a woman with her titties out being fed fruit. There's only one way to find out

do not watch this movie. Unless you want to
manifest the whole life. If you are trying to
manifest a new business or a new job, this
movie will completely change your life. Your
life will completely change when you understand the three things I'm about to tell you in
this video. Lesson number one, main character energy. Lesson number two, fear voting.
And number three, my favorite thing identity
shifting.

So I'm living proof that scripting is real.

Girlfriend, you're not the only one. Scripting
is real for me, too. So basically, if you want to
manifest whatever it is that you want to happen in your life. Fortunately for me, I didn't
have to spend a whole year for it to happen. It actually happened in four months. So
one day, I was in the office supply section at
Walmart. And I came across some notebooks
and I saw one that just like stood out at me.
And I was just like, I have to have this note-
book. And this is the one that I bought. And
of course, you know,

the John know that it's 100% illegal to try

to communicate with dolphins at your local nearby beach.

Did you know that if

you get caught doing this, you can get fined up to $100,000 or sent to jail for up to a year. Now, the video clip I'm about to show you is for entertainment purposes only I repeat entertainment purposes only. Traveling like me, there's a

new Mandela

Effect video

on the phone. Now we have FaceTime, Skype and other similar apps. But for the longest time, this was the future. I used to think oh man, when this is real well, no, we've made it because we're living in the future. But now that we have it, no one really wants to do it because he wants to get showered and cleaned up for a phone call. And the 1960s bell system later renamed at&t showed off their picture phone as a prototype at the New York City World's Fair. No, they didn't.

Course morning, John, you're looking well this morning. I'm very pleased and fried John, to participate with you this morning, this historic call.

But that wasn't the first showing of this technology. Even though at the time this was shown off as a brand new thing never before seen or realized. What if I told you that video phones existed and not just in concept but actually existed and were publicly used at seven years ago? Shit. Go ahead. Take a moment to take that in 87 years ago, Germany 1936 The world's first two way video phone service launched no this unbelievable historical achievement that seemingly spawned into existence was so far ahead of its time get this it connected to a range of 100 miles

to now cities in Germany.

So have you guys seen the experiment where the woman

talks into the water and asked the water do you know who I am and puts it in a pan and throws it in the freezer and the next morning the initial her initials was in the water

or we're gonna try that is

anybody even using search engines anymore because the way I stay on Tiktok searching everything and my thing is it can get even better. If

you're coming across this video it's not by accident. There's something in this reading that you need to hear. We have the 10 of Pentacles and the star card. You are so incredibly precious to your ancestors and this includes passed on loved ones from this lifetime okay, you're very very protected and you are so highly favored. Because you are the healer of your family bloodline you're resolving a lot of patterns you're transmuting a lot of pain they this as

it is for my superstar cell family and this iden-

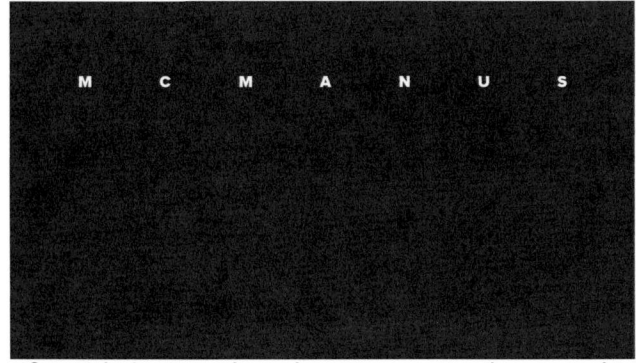

tify with being the chosen ones please take what resonates leave the rest this is my only account talk it's your

you're finding this message. It's for you. I don't give you are looking

for a bomb spiritual app. You're going to need this thank me later. It's called Moon lane.

We allows you to track the lunar cycles and astrology on the eastern calendar. It also comes with daily affirmations that you could post to your Instagram Stories affirmation.

This one was my favorite one. I am thankful for all the love in my life and I find it everywhere. If that is not self love. I don't know what self love is. And

I'm gonna prove to you right now how powerful I am and you are what our minds work together, I'm going to show you three things that you're going to see in the next few days. We're both going to see it. And we know that we're going to see it appear within our 3d In the next few days. So here it is.

When Brennan was sick, she was able to tell me that God sent him down to me, because

ANAMNESIS

God knew that I really wanted children. And
because he was such a kind, smart man,

the last time he told me, God sent me back
on a white horse and I woke up is Brennan.
invisible.

Call me crazy, if you want, but I've never like Storm. I don't know Susie, that might be crazier than the time that I went to go see a psychic for the first time. Now. This psychic was legit. She's one of the ones that was so good that she could help the police solve murders, find missing people, all the things. So I went to go see her and she was giving me my reading telling me things. Like I said, it's my first time. So I was taking everything with a grain of salt. I was like, okay, yeah, sure, fine, writing it down all things. But then she got to my youngest son, then she told me, your youngest son can see his spirit guide.

The story is 100% true. And I've got zero XSplit donation for it. Any of you that have followed me for a while or might have heard me talking about this one, six months ago, or something like that, you know that I'm not spiritual. I'm not religious. But what I'm about to share with you, I think ticks both of those boxes. So we're talking about a year ago now, and I know it was about a year ago, because it was my best friend's birthday. And his birthday is in August. And we decided to go out for a night out, went to a local village was called corn, just outside local, just outside Leicester. And it was dead. It was a Friday night, I was absolutely dead. And it's normally really busy.

It's normally a really, really good night out. So we decided actually, let's go to the local town. We've got the books in 40 or bloats, getting the books on a Friday night, how stupid am going to a local town found when these bars and it was really busy. And it was dead busy as much better. But it was, if you know what I mean, it was a bit of rough. And we were in there for a while. Now following up to this, I'd been obviously drinking awful, an awful lot. And I wasn't feeling too good. But I'd started meditating quite religiously. And I was doing the sort of couple of times a day, still doing it now. And we're in this bar, and the guys had gone off somewhere maybe to the toilet. So it's as that goes all the drink. So I've gone to the bar and ordered a drink, walk away from the bar and this bloke catches my eyes stood at the other side of the bar looks a little bit worse for wear. I don't know what drew my eye to him. But he saw me looking at him. And I walked over to the table where we were all sort of congregating around, put the drinks down and couldn't help but keep looking at this bloke keep staring at him. I got this really weird feeling in my stomach when I saw him. So now he's noticed me again and his mouth over what the fuck are you looking at sort of thing. Like I said it was quite a rough pub. And so I walked over to him. I said, Look,

Mater says I'm not here to cause any trouble says, you know, not starting any trouble is I want to keep staring that message. I don't know. He said, What do you mean, you don't know. I said, Well, I'm looking at you. I've got a really, really weird feeling. I've got a real sort of feeling of doom and doubt and whatnot. He's looked at me, he says, Well, looking at you. I said, um, nobody. Let me buy you a drink. So I bought this guy a drink. And we start chatting. And he's like, you know, what are you talking about? He says, you know, do you recognize you from somewhere? I said, No, I don't. I've never seen you before. as well. I've got a weird feeling that you're going to do something stupid tonight. And I said, I've just got the sense of a train or something. Will immediately the guy tears up, tears in his eyes. looks at me. Fucking hell are you? I said I'm nobody. He pulled out of his back pocket, a timetable for the local train station. Now where we live. It's a route straight to London and Leicester. But a lot of the trains don't go straight through, they stop at the train station. But some goes straight through. And he looked at me and he says, I was going to put myself in front of a train tonight. And he said, Yeah, look, here's the timetable. He said I knew exactly which train was going to go straight through the train station nonstop.

And that's when I was gonna do he says, I'll come down here to get picked up first to get some courage and I was gonna do it. I was like, fucking hell. So the guy starts crying. We both go outside, having, you know, a bit of a heart to heart. And he says to me, you know, what's gone so wrong in your life. I was probably mid 40s. Some reach to me. He says, Well, I think I'm gay. I said so far. He's 2023. You can do whatever you want. He said, Nice as a car. And he says, you know, the friendship group that I've got in the way that I've been brought up. He says that would be the worst thing ever. So we were chatting for a while. And he's like, Man, you've saved my life. He says, I'm not going to do it tonight. I said, No, don't. I said, you know, make sure that you sober up and you only think about things in the cold light of day. Once you've sobered up, you realize how much of a mistake you were going to make. So I gave him my phone number. And a couple of days later, you know, had a text message with him in it. You know, he's obviously still alive. But I've got no explanation for that whatsoever. Now when I say I had a feeling it was kind of, I don't know, just this feeling of doom and gloom, when you sort of looked at him, I don't know if it was body language that I was reading or whatever. But I've not heard anything in the local papers about anybody, you know, actually doing that. But he was, you know, he was distraught, bless him, and I hope that he's still there, but I'm sure there's some of you out there that are gonna say, yeah, that's, you know, spiritual and whatnot. Like I say, I don't believe in religion and all of those types of things, but something definitely went on there.

So I've seen a lot of people post this on brush here on Tik Tok, and I've seen a lot of girls do and I haven't seen a guy do it yet. And quite frankly, I'm gonna do

I'm sorry, what was happening next year. I'm sorry. Well, this absolute saw right here is Barbra Venga. You may have heard about her before. She is a Bulgarian mystic who has correctly predicted so many events through history. She was born in 1911 and died in 1996. And she has some insane predictions that have come true before. She has some for next year. So make sure you save this video and share it with someone to see if anybody's country. What does she got to say about next year, and some of it bit scary? First one, pretty obvious. I think everybody's predicting this in the moment, use the word power weapon deployment, meaning not necessarily

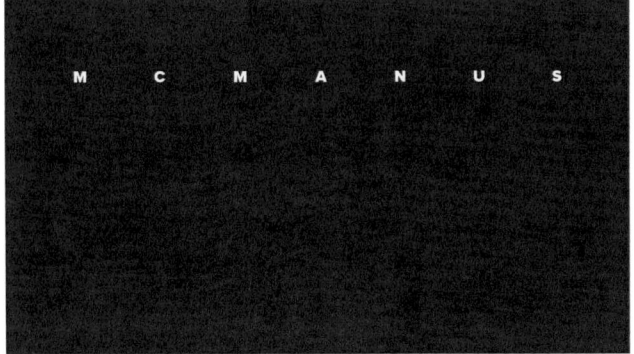

the end of the world, not necessarily a huge nuclear disaster. People begin testing them and firing them out and just playing around teasing the end of the world, maybe the end of the world. So 2025 prediction, you know, we'll look at this next time next year, it was set up as one I've been seeing a lot a load of people have been predicting this and she did as well, that there will be a new disease which will wipe across the world essentially, disease x, which is already hanging headlines. I hope this doesn't happen. But there you go. He said they will be designer humans next year. Not sure what to make of that is a bit odd. He also said there is ways we will be able to rewatch our dreams almost like a film.

Freeze river finally dries up and this has emerged something incredible has been uncovered. Something that had been hidden away. Beneath this flowing waters, or Thomas janai has delivered to hail.

You keep playing with God and you

leave this earth you are going to hail

Listen, in all seriousness, my last video was meant to have a little fun, make a light hearted joke, out of kind of depressing situation. I don't know what it's gonna take for black Americans specifically to recognize that this is a slave owners religion. And it in turn sound like a whole tip or nothing my sister, but the only reason why Christianity is popular with black Americans is because that was the first book. The first and for some of us the only book that we ever were allowed to read and was would be the first book that we would learn to read. To understand how to speak English, it is embedded in our culture, because it's a part of the trauma of our ancestry. Now, I'm not here to try to convert you to be a witch because I really don't give a fuck. This is not about saving souls. This is not about recruiting people or getting people to believe what I believe this is about my personal relationship with source for the divine with God, whoever it is that you believe is up there pulling the switches, my practice is entirely my practice. So

cool. Even the bag to get out.

Okay, somebody, please explain to me what is going on and then I'm not going crazy. Okay? If you know me in real life. Don't make fun of me for this.

One of the fastest ways to create something in your physical reality is to reverse the normal process by which we create things. Usually we create things first by thinking about I think I'll go to that party. And then we say something about it. As in Matilda. I'm coming to your party tonight. And then we do something about it as in showing up at the party. That's generally how we produce things in our reality. But if you want to really play tricks with the universe, and create magic, reverse the thought, word deed paradigm, turn it upside down and start with the deed. That is to say, act as if, if you want to experience abundance, be abundant, and do as abundance does

less than 60 seconds of your medium in disguise and don't even know number one you are afraid of the dark and you still are possibly number two, you're an empath you feel people's emotions and dogs. Number three, you have super vivid dreams and daydreams. Number four you hear people calling your name but there's nobody there or

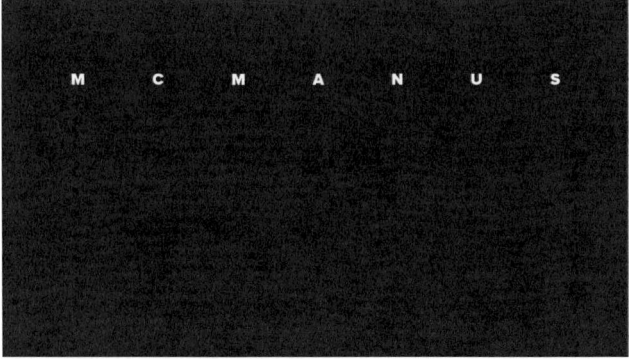

Okay, somebody, please explain to me what is going on and then I'm not going crazy. Okay? If you know me in real life. Don't make fun of me for this.

One of the fastest ways to create something in your physical reality is to reverse the normal process by which we create things. Usually we create things first by thinking about I think I'll go to that party. And then we say something about it. As in Matilda. I'm coming to your party tonight. And then we do something about it as in showing up at the party. That's generally how we produce things in our reality. But if you want to really play tricks with the universe, and create magic, reverse the thought, word deed paradigm, turn it upside down and start with the deed. That is to say, act as if, if you want to experience abundance, be abundant, and do as abundance does

less than 60 seconds of your medium in disguise and don't even know number one you are afraid of the dark and you still are possibly number two, you're an empath you feel people's emotions and dogs. Number three, you have super vivid dreams and daydreams. Number four you hear people calling your name but there's nobody there or speaking. Number five you have issues with your electronics they break down on you or they malfunction. Number six when you were little you've seen spirit before. And orbs are you see things out of the corner of your eye. Number seven when you're around people in pain that you sometimes can feel their physical pain, such a crazy phenomenon. Number eight things aesthetically have to feel good around you. I know if you ever add you have a mess, but still you have to have nice things it has to feel good. Number nine you smell smoke or other things like cologne around and no one else can smell it. Number 10. You were always known as being super intuitive. You just know things you have a gut feeling. Number 11 you're drawn to mediumship and spirituality. Before

you were born here, you get to pick your opportunities and what your life experiences will be out on the screens through the help of our council of elders. These life between life sessions have also revealed that we go through somewhat of a dress rehearsal with our family members and friends before incarnated. This involves meeting with our parents in this world, even before you're born and giving ourselves unique challenges. We learn from life between life cases that older souls incarnate less, and choose to stay back and become spirit guides, and take on other duties in between lives. And while younger souls might incarnate more often, for faster or easier assignments, many souls choose not to come back to Earth at all. Because time does not exist on the spirit side. And it's nonlinear. Everything is happening in simultaneity. And linear reality is only an illusion created.

Noah is actually a term that was given as a biblical term, he had a couple of names in history that we actually were was called by his name was own tend to pitch them, or his name was via sudra. I like no a better his story is is really quite amazing because he was actually the very last of these pre diluvian kings of one of these original five cities of Sumur, known as shuru pack and went

to the hardware store and I bought a roll of metal duct tape. And I just taped it across the bed, I threw a wire out the window, and it had a ground rod outside like connected to the ground rod on one side and connected to the metal duct tape that I had laid on the bed. So when I laid down on the duct tape, I was like, grounded, because it was connected to the earth. And I woke up the next morning. And

I thought, holy cow, there's something going on here. Because normally for me to go to sleep, I had to take Advil and went down to University of Arizona and one of their medical libraries. And it was nothing. In fact, I even tried to find the cause of chronic pain. And they didn't know nobody knew. So I ended up having to put together my own study

underneath the Pyramid of the Sun and also great preeminent at Giza, they have the ability to absorb something called Physiol static electricity from running water.

So most people that when they wake up in the morning, they're a clean slate, they're not thinking anything, they're not feeling anything in a matter of moments. Because the brain is a record of the past, they start thinking about those problems, those problems in their life, that are associated with certain people, and certain objects and certain things at certain times and places. The moment they start thinking about those problems, they're thinking in the past, because they are memories from the past expression of proteins is the expression of life. Because proteins are responsible for the structure and the function of your body. In order for a cell to make proteins, a gene has to be stimulated or regulated. So what are genes that think about genes as a library of potentials. And so there was a myth after the human genome was discovered that genes create disease. But if you study the research, less than 1% of people on the planet, are born with genetic conditions, early onset genetic conditions, like Tay Sachs disease, sickle cell anemia, type one diabetes, the other 99% is created from lifestyle, behavior and choices. Sit down, close your eyes, slow your breath down, place your attention in your heart. That's where the your elevated emotions start.

The human being is a really interesting complex in that probably the hardest thing for us to be able to access is ourselves to get to that really fine fine, beautiful highly intelligent, highly creative aspect of self that continues beyond physical death. That part some called the soul Some call it spirit soul development, whatever you want to call it is much harder for us to be in direct contact with than it is for these other beings to kind of intercede on our own behalf. The sole will have a directive for example, but then it will take a GO button mean of sorts that can reach into this dense, conscious mind complex of ours, knock us on the head put in a vision or a word or feeling so that we can then hear the message of

our own soul.

While I'm reminded of what Socrates called his demon. That was the concept of this messenger. This being that was essentially an entity that was with him that gave him a ration and inside his personal Genie Genius. But it also might be a gin. Exactly.

It could be a gin which of course, we know we'd be more toward the demonic realm in some cases in terms of

So what is an energy field? You are an energy field. What is a morphic field? It is an energy field. What is collective consciousness? It's an energy field. The difference is, the human energy field has a physicality to it. What does the energetic field look like? The easiest way I've come to define it, and see it is as a bubble, a human energy bubble,

stem cells from the child actually migrate in to the maternal system. But studies in Rome at the Vatican guess what they found? Many of the child's stem cells end up in the nervous system of the mother. So think about it this way. The brain of the mother is like a culture dish that contains stem cells from her own child. And they say so what's the significance

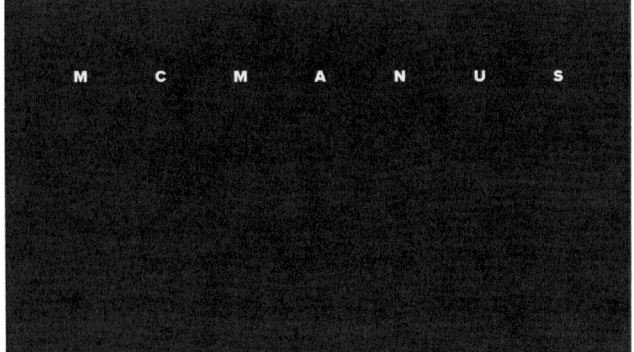

I say, her child can be on the other side of the world have an accident, and she can experience it.

Pluto is power Libras aesthetics, beauty, balance, fairness also. So their superpower that generation superpower is their ability to make everything beautiful, timber, the

most fundamental level, we keep trying to recreate our core traumas, casting the people we meet in our life into roles in our own psychodrama. The biggest rules for us to cast in our subconscious psychodrama, are the bad mommy and bad daddy. Once we have given someone that rule, of course, without informing them of this, we can then form a relationship with them, and then fight with them to endlessly recreate our childhood traumas, with a subconscious hope that maybe this time it will turn out differently, and will now get what it was we needed before.

So let's talk about resistance training, you

have to know that you are creating every
single thing in the universe.

In your mind, we learn from the income if I
have your you have

psychic abilities and just haven't unlocked them yet. So here's how everybody's born with psychic abilities, but some are more and more aware of them than others. And the more aware of your gifts you are, the more powerful you become. Awareness is truly key to unlocking managing and strengthening your psychic abilities. So today I'm going to teach you how to unlock and strengthen yours with a part of my hyper awareness strategy that is meant to help you tap into and strengthen your psychic abilities. Right now if you focus on what you hear, you'll start picking up on noises you didn't before. And if you put all your focus on your nostril you start picking up a site you didn't before becoming aware of his psychic abilities is the same thing. Make sure to screenshot the screen and after this video to the exercise I'm about to tell you sit down with your eyes open and stare at the Red Dot and we want to focus on the present moment using your breath to be more present. Now you want to notice these things what are your five senses pick up your feelings and sensations Do you feel throughout your body? Are you starting to see small glimpses of energy orbs flying around? Stay here frequencies you didn't before? What do you feel around you after this video, do this exercise for a minimum of 10 minutes to see

what you become aware of you will see a huge difference of what you perceive just I'm using this part of my hyper awareness strategy. And if you want to reach your highest spiritual potential, check out the top of the page for more help.

These guys were driving on the countryside at night when they caught this on camera. Pay attention to the windows when they start filming the old houses.

When you look at the clock, you're seeing 1111 What the hell does this mean? When 11 is duplicated as it is with 1111? All of these associations that are formed with number 11 are amplified. 1111 is the ultimate confirmation number. It's essentially a universal Yes. When you see 1111 Is the universe communicating with you in a way where it's saying yes, what you're thinking right now what you're saying right now or what you're doing right now is definitely confirmed by the universe at large. You are on the path, the path that you have decided on before coming into this life, the path that's in alignment with your purpose, the path that's in line with your desires. Best experiencing

the pyramid first of all, when you stand next

to the Great Pyramid time slows down because it's time dilation bubble, they knew all this stuff. I mean, the Great Pyramid itself is a gigantic multifunctional stone computer and power generator. It's got multiple functions. And so when you begin to analyze it as a technological device, not as a module meant all of a sudden all these question marks start popping up. And when you begin to start looking for answers to those question marks and start studying engineering books, you start to see that we're looking at a real machine, a real machine. And it's so ingenious how they built this, there's a couple of ways that it can happen. First of all, I never was on my ramp mud ramps is that this is a complete farce because you would need more mass in the pyramid in there you can

splash water drip, Splash Splash drip waters.

This person's glitch in the Matrix story has honestly rattled me to my core, and I need everybody to see it. Because I need to know what people think. And I need to know what the answer to this is. Laura posted this story in my Facebook group stay on,

I honestly have not been able to stop thinking about it. It's just rattling around in my brain. Please help. So Laura is deeply in her bathroom when she found her ring in the place that she always leaves her ring when she is taking a bath, except that she was already wearing her ring. The ring, what she found was exactly the same as the ring that she was wearing. The same size, the same shape the whole lot. Except the really she was wearing she'd only ever bought once she has had the same one for years. rarely takes it off. When she does take it off, she puts it back on again, she had lasted I think it happened to it. Except now she has two of them. But it's not just that, okay, the ring, as you can see here has a bend in its clasp, where lower caught it on a jumper years ago, it got stuck, and it bent the class. And the new ring, the parallel universe ring has a bend in the clasp, but in like a mirror image. So you can see, this ring is real. This isn't a fake. This isn't some sort of Photoshop, there is the ring, the new one,

there's the one that she's been wearing the whole time.

What would happen if you were to contract the intrinsic muscles in your body, the muscles you use for elimination, the muscles you use for intercourse, you were to contract your lower abdomen, pull your belly button in towards your spine and contract your upper abdomen, the act of contracting these intrinsic muscles in the ancient yogic tradition called the BONDUS. As you squeeze these muscles, you begin to push that cerebrospinal fluid up. Now what if you were to take one slow, steady breath. And as you slowly follow your breath, and you can track those intrinsic muscles as you follow your breath and you bring that breath all the way up to that pineal gland. And you hold your breath and you keep your attention on that pineal gland. That's your target. The act of contracting those muscles, along with the holding of your breath, would begin to push cerebral spinal fluid right up into your brain. And as you begin to hold your breath, you're pushing that cerebral spinal fluid, right up against the crystals of the pineal gland and applying a mechanical stress and then the mechanical stress begins to electrically activate those crystals. And when those crystals become

MCMANUS

electrically activated, we are turning on the radio receiver in the brain. The songs you sing out loud actually manifests.

This is true. This is why I like to write high vibrational music.

I wasn't the only one who practices

Have you have you practiced in going out
and picking up people off the street the

easiest way?

Tell him and why.

Please let him speak showing paid by
Bianca okay was a was last week hides my
expenses, insane

belief in my own ability to manifest things.

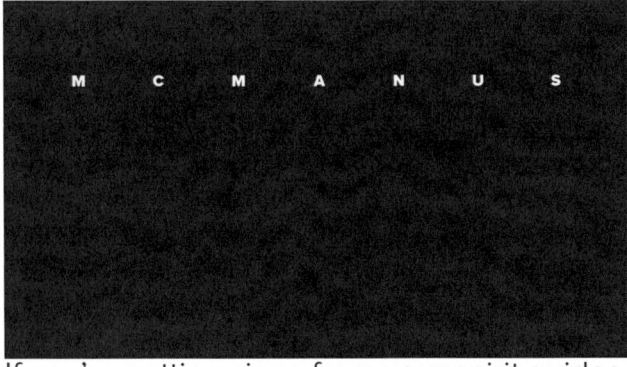

If you're getting signs from your spirit guides, like

repeating numbers or other synchronicities. This is how you can get the message they're trying to get to. The first thing that you have to do is stop trying to look up the meaning of the signs because all you're doing is reading someone else's interpretation of what those signs mean. You want to get the exact meaning for you. So do this, treat the signs as if your spirit guides are knocking at your door trying to get your attention when you notice one of these signs, bringing awareness to your immediate environment. Just sense and feel the energy that's there.

Because in ninth, I think it's on tape in 1998. For example, in 1998 at an event he said that there would be a terrorist attack in New York before the end of 2001. And it was unlikely to change. And here we go. He's

about to do it. And my daughter who was 18 She said not yet and he said he's been gone the entire date is not that we can do he's already gone. She said no, he's still in there. Don't care he's still he's still in there. She said why not just wait extra days so they waited a went to the house started packing up my home started making Memorial Plans my job had drafted up the condolences letter already for my family and a family with his cleaning out the house until the next morning at nighttime. The doctors called them and said you won't believe this come back to the hospital and we found a faint pulse. So many

years ago, when the government realized what alpha was, they made it illegal to study it all over the whole world. Because it because it makes people superhuman alpha, by the way, what it is one single word is joy, alpha is essential for higher spirituality, higher consciousness, you do not have a choice in this, you have to be able to fire alpha at a certain level and and relatively balanced across the four quadrants. If you can't, you can't proceed is just the way it is. Children are filled with it. Their babies, just one big elephant machine. And teenagers to follow it through there, they've got lots of alpha being fired out. But as you get older, the alpha starts dropping and 25 and 30. And by the time you're 40, and 50 is way down by 20 and your 70s and 80s. The alphas just probably gone away, and and they start entering into theta, and eventually you enter into delta and you die. If you're old, you're going to learn how to redo it and change your age

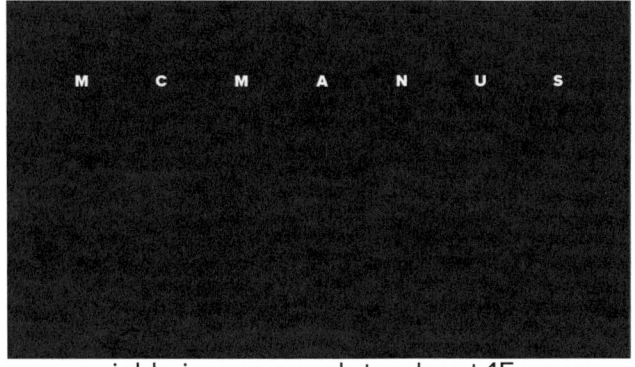

very quickly in one week to about 15 years.

So if you're thinking thoughts that are connected to emotions, like anger, or frustration, or pain or suffering or guilt, those thoughts produce a slower frequency and be try

making all the bad decisions. And so it is with the battery packs in the human body. On one side, it's how many electrons are we putting in and there are a lot of electron donors. Electron donors can be getting out in the sun can be walking barefoot in the grass, it can be leaning up your body up against a tree because the tree has electrons will donate them to you. Hugging an animal you know common way that makes people in nursing homes feel better is by them a dog or a cat. They hold the dog or the cat and get electrons from it. The animal runs outside and recharges itself and bring some back to you. Moving water and this is electron donor whereas still water tends to be electron stealer moving backward

another glitch in the Matrix story that I found this one is short and sweet. So let's get into it.

This video is so disturbing the number of people have felt ill after watching it. Proceed with caution. One here smells

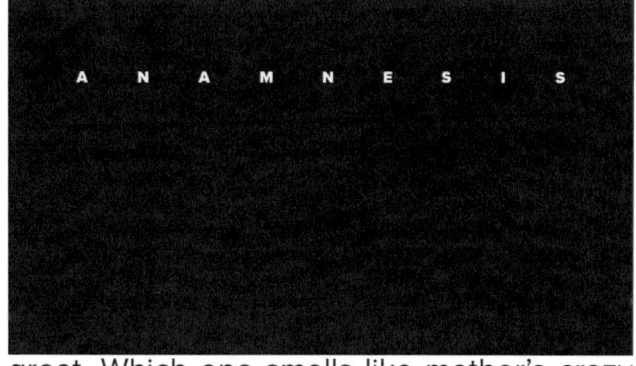

great. Which one smells like mother's crazy sister. Okay, two things so it smells so good. She could have been that crazy. I don't think so. I think so. Well she put her poodle one time in a microwave oven to eat it to eat it oh no no no now now now now

it exploded and they were both found dead she must have been out

this video is so disturbing the number of people have felt ill after watching it.

You want to sound smart don't say your hard work and use the more advanced word asked to do a sound so you're a scaredy cat use the more advanced word puzzle animus you can enhance your vocabulary everyday with words like this by using the vocabulary and even put it right on your lockscreen and learn new words. Every time you glance at your phone Catholic girls.

If a guy says to you, I'm literally just gonna like full on outright tell you this. If a guy says I'm literally only in this for sex, he's literally never going to change his mind. Okay? Because what we love to do as girls is be like, Oh my god, me too. It's just like, you know, it's just fun, but like in the back of our head,

it's like wedding bills. Think I know the dressing wearing. We cannot be like this anymore. Okay.

Did you know that the law of conservation of energy is hidden within our multiplication table, and you only need to use basic math to see it. This

man found a $20 bill with a watermark of a horn, George Bush.

I just believe that they need to get out of the Ohio Valley and let others know about it. When it is in his religious Bible I have had for a long time is a $20 bill and mostly $20 bills which when this came out in 2004. The watermark is Jackson and Jackson. day when I got this bill, we were all looking at the light because it was a brand new bill and as we were looking at it with Jackson Jackson Very solid bill. And I have a lot of friends that you know, witnesses bill at the time when they got it. That evening, I took it home and put it in my angel book, why was a bookmarker because I didn't have my other angel book marker. And I basically just tuck it. I stuck it in the book and what sleeve I thought nothing of it. So the next morning when we get up to you that $20 bill for gas. And as I did, I couldn't believe what I saw when I looked at the light the final time, when I looked at the watermark. It looked more instead of Jacksonville, you looked at his Bush, if

you're into astrology and understanding yourself better vibes takes it to the next level. It's all about diving into astrology and science, getting to know yourself on a deeper level and connecting with people who share similar interests. The best part is that it's completely user friendly and free to download on the App Store.

This tick tock page is probably available, beautiful woman that turned into an ocelot She lived with her husband and she had a few kids, right. She was beautiful. Like imagine like the most beautiful woman in the Philippines, you would never expect her to turn into something so like crazy. And I mean, and one day she would come home from her work, and she would start treating her kids like super, like, Oh, you have to go get this you have to get the chores done. Do that. Yeah, and her husband's watching isn't like no, like, this doesn't seem weird, but homiletical right. And then she starts being fun towards the kids love the husband is out, right. And the husband usually comes back from

work and he's hungry, so they don't have a lot of money. And what she does is the whole urban legend is that she was possessed by a demon that made her crave human flesh. You

can't actually sell your soul. But I'll explain what it means when celebrities sell their soul. We are all part of one unconscious as you are me I am you you're also everything and everything is also you Everything is gone from the trees to the skies to the plan is to the insects to the animals to you. So if someone were to actually sell their soul, that would mean they're giving all of us away. And that's

monumental difference between these two, let's talk about it. This one is just coconut and peppermint. Say that coconut and peppermint. This is more for teeth whitening and minimal results. If you have healthy teeth, you can use this one. If you have bleeding gums, you want to use the Mickey D version. Because of all these wonderful added ingredients, spearmint, peppermint, cardamom, fennel, tea tree, Clover, oregano, vitamin E, D, and K two for bone health. So it's all what's in the sauce. Look at those ingredients. Compared to these ingredients. This is the super viral one on Tik Tok, which is the Mickey D version. This is the one that I recommend.

And what people don't understand is the Sun is a portal.

So I'll just tell you this example

I had this client come in, and she wanted to lose weight. And she tried so many different things. I mean, literally, she went on a juice fast and the weight went in budge. So obviously there was something going on. And usually when a client wants to lose weight, their subconscious will say, Oh, the weight acts as like protection, you know, that's usually the case. And if you can figure out a way to feel protected, the weight will come off very easily. And the subconscious needs to understand this, and so does the conscious mind. Well, this wasn't the case, the higher consciousness said she needs to go outside without our contacts. And so that was a little surprising.

I asked, What do you mean, and they said, she needs to be exposed to unfiltered sunlight. Because humans, all humans need a little unfiltered sunlight because ultimately, what's available to humanity right now is the power of this intelligence coming from the portals of the sun. It goes through the unfiltered eyes straight to the pineal gland, and it unlocks the human power.

So true. And that's interesting that came through and you and I did a session that you've transcribed a little bit about in the sun played hugely Yeah. Back in the Atlantean times too. And also the whole notion of going out in the morning for example for people who are just having having this memory come back again to go in early blush of the morning when the sun's first rising and even if you're just looking at the sun through your eyelashes to begin with, to just open your eyes for however many seconds and take the fullness of the cool Sun's information into your being first thing in the morning. has a phenomenal I'm so glad you brought it up because I stopped doing it too. has a phenomenal effect. But they're saying also go out in and I would think bathe yourself in it however you can but they say definitely through the eyeballs unfiltered is

so interesting because I'm always wearing my glasses. So I stopped doing that even just a little bit. They said, Yes, unlock these dormant potentials within ourselves. And what

if a person would sit down and let their body relax, totally relax, and then start to visualize in their mind, see themselves the way they want to see themselves. See how you'd like to live your life, see yourself living it that way. Now understand that that's a picture in your mind. When you pick up a book, the book

is nothing but a picture that an author has painted in words. Van Gogh, the great artist was asked one time how he did such beautiful work. He said, I dream my painting. And then I paint my dream to get the picture in the mind, then painted on the wall, or on the canvas? Well, if we would relax and build the image in our mind, of how we'd like to see ourselves, and then take that picture and describe it, write it out. In the present tense, I am so happy now that I see myself and write it out. A lot of people will laugh at this and say, it doesn't make any sense. It makes a lot of sense. They can't tell you what doesn't. Okay, right. Hi.

I love you so much. Okay, so we're gonna do a couple questions that people really want to know, because you've said things that have triggered curiosity within them. So number one, you said that humans are not the main species, so long not wanna know. What's mean species? There

is none. There basically kind of is not an angel is just can be a species of its own depending on what type of species

so there's multiple species of angels. There's even so would that be considered if we had

to pick and say So angels there? There we go. Angels are the main species lightspace, lightspace, lightspace, and Darkspace. Can you explain to everyone what that means?

lightspace is basically like basically, like I'd say heaven, and it basically is like dark space but different in with higher vibrations. A dark space is one with a bit lower vibrations and a bit of normal type of vibration Number Use

the bureau and Anunnaki here's what

I have another absolutely fantastic glitch in the Matrix story for you and I need to share it with you and I need to know your thoughts about what the heck happened to this person. For context. Amy sent me the story months ago and I had forgotten about it until somebody commented on it today. And this is the podcast episode The story was right out in the sun Statue

of Liberty seems to be getting weirder and weirder. People have found that there seems to be something huge lurking under its pedestal is

footage of CCTV footage from a grocery store. And it's said to be of a man who's been plagued by the Grim Reaper himself.

As you can see this man is looking through his wallet everything seems fine. When all of a sudden something really weird just happens. Just watching tell me what you think.

This just might be the oldest personal life.

The avoidant is in control of the relationship and some of you who are on this end of the spectrum, you know how horrible this feels. The relationship is always controlled by the person who can pull away, you

never actually sleep you just go to a different reality. They just light and call the dreaming so you stay trapped in the physical realm though,

I just had the creepiest experience of my life. It's this app called Da Vinci AI. When you go in and write your name, the month you were born in, and your city of birth. It's supposed to keep you elements from your past life. Every time you try. It reveals another part. Like the places you've been the people you were with. My seventh try was shocking. I still have the chills. You got to try DaVinci right now.

I've been hearing about throat cancer, a lot of colon cancer. Yay. For right now. Suffering

from Call it cast. I've been trying to get him in contact with you do you busy? Yeah, but ain't too busy for you. So So yeah, you busy like he's he's going through chemo right now. But,

you know, they're saying that I believe it's HPV is what's causing a lot of throat cancers and colon cancers and things like that. Like,

I mean, well, so I mean, I'm pretty sure a lot of your audience know now but my perception and how I look at disease is super, super different. So when I see cancer, or the mutation of sales or malignant cells, going through a mutation phase, I think they can all disease to me as the body trying to heal itself. I'm actually writing a book right now that will be out in July. It's called Healing the illusion we call disease. Okay. And basically, the book is taken every disease known to man and shifting their perception about what disease is, and my stance was trying to come out. I know. It's hard for them to come after me because I don't believe in disease. So how can you sue me you're trying to send me away for healing disease that I don't believe in. So now, so that's why it's so hard for them. They really even get me in court. Because my position and my stance is there is no such thing as

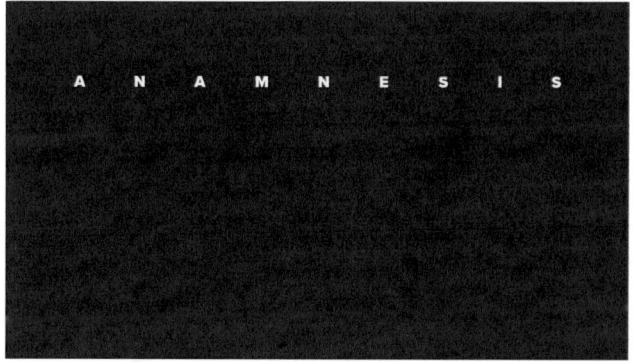

disease. Disease is the body creating symptom ologies to heal itself,

you'll most likely be dead in 11 months. Why is nobody talking about this? A few days ago, NASA just made a horrifying discovery during a night sky observation session, something that might signal the end of life on Earth as we know it. Just three days ago in Washington, NASA launched a telescope into space called the B 22x. This telescope was specifically designed to study the movement of meteorites and asteroids in space to predict potential collisions with the Earth. On September 27, a B 22x identified something horrifying, it calculated the trajectory of an asteroid set to hit Earth in 11 months. The impact date is projected to be September 2 2020. For astronomers and scientists estimate there's a 79% chance the asteroid will strike the Earth's surface. The asteroid size is four miles in diameter, and moving at a speed of 10 million light years per second. And

if you're single and you're every time you get a warning, you have the ability to manifest anything into existence. When you get worn, you have accumulated sexual energy and your sacral that wants to come out and be released. And the sacral chakra produc-

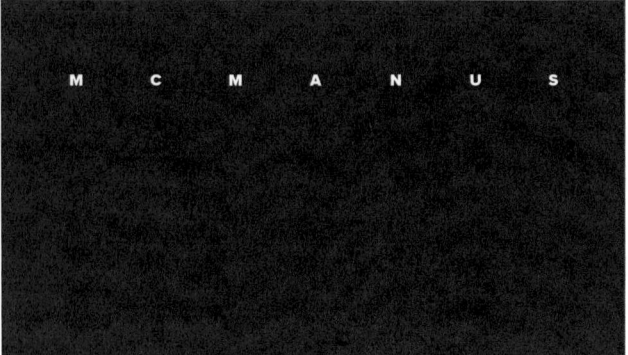

es this sexual energy to allow self expression and creation of not just human beings through your entire life is a form of creation. This means that you can give birth to ideas to businesses to art to new invention to whatever you desire using this sexual energy. And this is what I tell my clients to do with a spicy energy you can use this energy to help you get creative for any work that you're doing to bring your ideas into reality or you can self pleasure and use his energy to manifest your desires by just feeling and visualizing this dream outcome you desire while you go to town with yourself and for this reason self pleasure is very sacred so make sure to treat it like it is truly one of the most impressive and useful tools human beings have to create what are your thoughts on this

guy and then

I got that on camera I've got that I just went
that way no life

ANAMNESIS

yes, I know that I'm high right now but look at this lady she's hitting a tennis ball against absolutely no circumstance just bouncing back to her like this

think I started fasting but I'm getting thin. I eat just watermelon and bananas and other so this is the thing that has to be a lifestyle and you have to tell yourself that you're going to get strong a lot of it.

We're driving past what we call a what is it called? Yes, but people called graveyard

I have with my New York City family of six. This is

true consciousness. You are one of these three.

What if I told you that every single day of your life, you've been practicing manifestation those who own

dogs, you know that. Every time you come back from wherever you went. The dog is happy and it's jumpy and it wants to lick you in the face. Even if you just went to get mail from the mailbox I

named L'Oreal

Voluminous,

M C M A N U S

like 10 years ago, I got chased out of my family's woods by something that I think was Bigfoot. I grew up on about 100 acres of land in the Appalachian Mountains. So we had the farmland, and then we had three mountains. And I know every inch of the land like the back of my hand, if you walk to the clifftop on the mountain behind my mom's house, it opens up to like this beautiful view of the valley below, it takes like two hours to walk there. And there are caves that I've never been brave enough to go in them. But my aunt said that the last time she had walked up there, personally, she heard chanting coming from the caves that did not steer me away. For privacy reasons, I'm not gonna be using my friend's real name, we'll call her Jenny. So my engineer decided we were going to walk to the cliff, it was a

little month ago, I was talking about how the Vatican scares the absolute shit out of me. And so I started reading a real PDF document called the secrets of the Vatican. And I'm going to tell you five things that the Vatican is hiding that no one knows about. I will warn you, some of these secrets are very eerie, and they actually will have you thinking about what's going on present day, as well as what is to come. So if you're someone who scares

easily just grow secret number one, the Bible actually contained 80 chapters. And now the Bible has 66. Watson, what is time?

Do I have to answer? It Right, this gets very deep very quickly. So if you're not inclined to follow it, I suggest that you don't watch any further. So for most people, we watch the hand of a clock, say, moving around, maybe we watch it for 30 seconds. And we say 30 seconds has passed. Well know what you've really done is you've watched the motion of a hand on a clock. The notion that is 30 seconds is something that we invent.

I don't know if you guys have noticed, but things are getting real weird. And now I'm not just talking about war, and all the crazy stuff that's going on in the world. Today. I'm talking about like actual literal things that are happening to us, as human beings, such as technological glitches, you may have been noticing it that your technology has not been working properly. That over this past year, it's almost kind of like mercury retrograde on crack, like nothing ever works. And the internet has been super, super shiny. And I've been hearing this from everybody got away. Not to mention time dilation. This is a big one this year. Now granted, the time dilation really started up in 2020, people started noticing that we were losing a sense of time, things were either going like really fast or really slow. And it wasn't just like just casually being Oh, hey, you know, time is flying guys. It was like no a collective amount of people were agreeing that something is off at the time. And may have noticed that 2023 has also flown. People are losing track of time right now.

Okay, that it becomes person

okay.

Okay you guys love me with the queen?

Who is conducting some of the first experiments with pure THC on human beings.

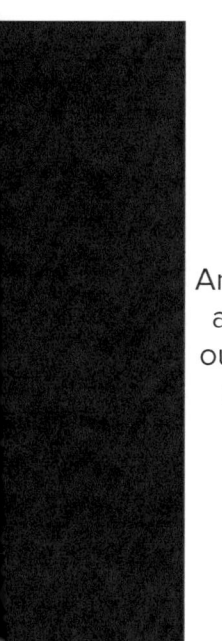

And by mid 2025. Now again, I had no idea about this. I didn't talk about this in previous videos because I never went back and read these books that I'd written. Right.

And so we're in poltergeist spoof

over here buying some time worth it for
me.

I hate to break it to you but your memories can work.

But what about Matthew Perry? Oh, what's the theory on that his whole day was actually planned. You go on his Instagram I don't know if you've seen it but he starts posting like this random videos. It's such a bummer and it shows Batman Hold on. And he goes to the ceiling shows the bad symbol. Right Okay. Usually in the cartoons What does a bad symbol symbolizes someone's in danger or

so I heard you had a near death experience where you were dead for over an hour. Can you please share that story with yes,

your antibody bags were

right on the diet about 45 minutes but I was likely dead well over an hour total. So they think I was dead for at least 30 to 45 minutes before they found me. But I was cold like cold to the touch. Back in the day, I was an amateur bodybuilder I was taking a fairly new supplement. Turns out the supplement was toxic and I ended up aspirating in a public bathroom so I passed out started to fall My aspirin on that vomit and ended up dying right there on the bathroom floor. And then out of nowhere, I felt like I was sitting in a very comfortable movie chair. And I was

watching this movie, but the movie was this scene of this body on the ground. And I was looking at everything from above. But what's weird is it didn't feel like it was me at all. Even though I was sitting there looking at my own dead body, I couldn't recognize it. It would almost be like going to a movie like real movie and seeing someone dress like you and looking like you on the movie, but you're like, that's not me is me is over here watching the movie.

My HOA president called the cops on me for being too loud. So I bought his shed and Atlas. Sheds land pays me rent every single second of the day for the rest of my life.

This happens at an armory in San Francisco. The search of security guard caught this on camera, and he said that he has a lot of paranormal experiences working this shift and this first time he's actually caught something too.

I need you all to watch this terrifying video and then I'll explain what happened.

As far as I can tell, it's just about letting the universe know what you want and working toward it while letting go of how it comes to

ANAMNESIS

pass. So if you have

a square glow shape, this is what it's going to look like. You are going to focus on the minimis and if

MCMANUS

you haven't seen that video, I advise you to go look at that video. Right now. I have the juiciest square shape and I've been trying her methods for about a week now realist

conspiracy theorist breaks down some strange urge you guys check this out? We got the sun the rainbow around the sun.

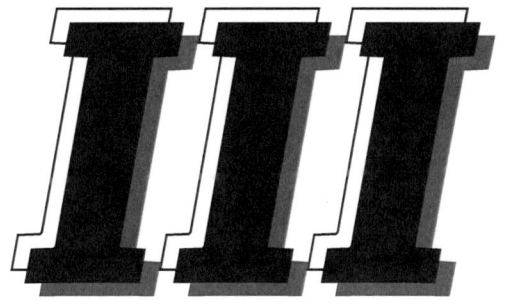

INTERVIEW

During this phase, I put the transcript into conversation with ChatGPT. I've broken up the transcript up into sections, based on time and theme, and asked ChatGPT questions about certain representative selections.

To prepare for the interview, I wrote out a series of questions centered around the following themes: **authenticity, integrity, rationality and correctness.**

But as with any interview, it is impossible to totally prepare. The text harvested from TikTok was unforeseeable and subject to algorithmic divination. And like any interview, when face-to-face with the subject—in this case ChatGPT—my questions did not follow a rote sequence. Rather, the process involved listening to what the subject said, and following up with new questions.

This unpredictability is always part of the interview process, and raises questions about the role of listening, connection and positionality (power) in the journalistic process.

The following interview was conducted with ChatGPT 3.5 and ChatGPT 4 on 11/29/2023.

You

Please tell me what is going on in the following dialogue:

"Oh God

no one no one over here what are they

on the side of the boat

what are those trails fast I mean that water some swimming real fast I need that water yeah

noise What is that noise

I heard it I heard it I hear it

oh my god"

ChatGPT

The dialogue you've provided suggests a scene of heightened tension and urgency, typically found in situations involving an unexpected or alarming discovery. Here's a breakdown of the possible context and emotions conveyed in each line:

"Oh God": This exclamation often denotes a sudden realization or shock, suggesting that the speaker is encountering something unexpected or alarming.

"no one no one over here what are they": This indicates confusion or a search for someone or something. The speaker seems to be looking for others, possibly indicating a situation where they expected to find people but didn't.

"on the side of the boat": This locates the scene on a boat, adding a maritime context to the situation. It could imply that the speaker is looking over the side of the boat, observing something in the water.

"what are those trails fast I mean that water some swimming real fast I need that water yeah": This suggests the speaker is observing fast-moving trails in the water, possibly created by swimming creatures or objects. The urgency in their tone could indicate a sense of danger or the need to act quickly.

"noise What is that noise": The mention of an unidentified noise adds to the sense of uncertainty and potential threat. It could be something related to the water, the boat, or an unseen entity.

"I heard it I heard it I hear it": Repeating phrases like this conveys a growing sense of urgency or fear. It suggests the speaker is trying to confirm the presence of the noise, possibly to understand the situation better or alert others.

"oh my god": This phrase, often used to express shock or disbelief, reinforces the idea that the speaker is witnessing something extraordinary, frightening, or overwhelming.

Overall, this dialogue seems to depict a tense and possibly dangerous situation on a boat, with the speaker reacting to something they see in the water and an unidentified noise, indicating a sense of fear, urgency, and confusion.

You

Thanks for that. I see you've labeled this new conversation "Tense Boat Discovery" but it doesn't sound to me like the object causing distress or being discovered is a boat. Did you think it was?

ChatGPT

Upon re-evaluating the dialogue, it appears that the setting is indeed on a boat, but the source of tension or discovery isn't the boat itself. Rather, the speaker is on the boat and is observing or reacting to something else, possibly in the water or in their immediate environment. The references to looking over the side of the boat, seeing trails in the water, and hearing an unidentified noise suggest that the speaker is reacting to an external element, not the boat itself. This could be something in the water, an approaching object or creature, or an unexpected event related to their maritime surroundings.

You

Right. It's interesting to me how you made the boat the object of discovery there. That makes it sound more straightforward than the text suggests. And that leads me to my next question: do you think that something must be observable in order for it to be true?

ChatGPT

Your question touches on a deep philosophical issue concerning the nature of truth and observation. The idea that something must be observable to be considered true is closely related to empirical and positivist

philosophies, which emphasize the importance of observable evidence in establishing facts and truths.

However, many philosophical perspectives
argue that not all truths necessarily require
direct observation. Here are a few key
points to consider:

Empirical Evidence and Science:
In science, empirical observation is
crucial for verifying hypotheses and
establishing facts. However, even in
science, there are truths inferred from
indirect evidence or theoretical models (like the existence of atoms, black
holes, or dark matter) that cannot be
observed directly.

Abstract and Mathematical Truths:
In mathematics and logic, truths are
often established through reasoning
and proofs, independent of physical
or observable evidence. For instance,
mathematical truths like the Pythago-
rean theorem are considered valid regardless of empirical observation.

Subjective Experiences and Qual-
ia: Subjective experiences, such as
emotions, thoughts, or consciousness,
are true to the individual experiencing
them but may not be directly observable by others. The reality of these experiences does not depend on external verification.

Historical and Testimonial Truths:
Some truths are based on historical
records or testimonies, which may not
be directly observable now but are
considered true based on the credibility of the sources and supporting evidence.

Philosophical Theories of Truth:
There are various theories of truth in

philosophy, such as the correspondence theory (truth is what corre-
sponds to reality), the coherence theory (truth is what is coherent within a
set of beliefs), and the pragmatic theory (truth is what works or is useful).
Each of these approaches has different views on the role of observation in
establishing truth.

In summary, while observable evidence is
a critical component of determining truth,
especially in empirical sciences, there are
many aspects of truth that go beyond direct
observation. Truth can also be understood
in terms of logical consistency, subjective
experience, historical documentation, and
philosophical reasoning.

You

Ok. What you've shared here makes a kind
of sense. I'm interested in how you've cit-
ed your points in a list format--that almost
makes it seem more quantifiable. But I'd like
to move on to my next question. If I give you
a portion of text, could you distill the major
themes and the tone of the text?

ChatGPT

Absolutely, I can assist with that. Please provide the portion of text you have in mind,
and I will analyze it to identify the major
themes and the tone.

You

this is your brain on New Age supremacy.
You can find its foundational beliefs in like
grandiosity. narcissism, like solipsism. And

yes, you guessed it supremacy, you can identify its adherence by their willingness to classify their fellow human as being inhuman as being background characters as being soulless husks or NPCs that serve the sole purpose of making their life look complete, because they are in fact the main character. These real people will often identify themselves by other labels such as Lightworker, or indigo child, or like a star seed. Those kinds of times are often the thing that they will claim to be an addition. They will also suggest that they are here to spread light, love, healing harmony, bring the world into a new vibrational pattern of love and happiness and joy. But never, ever, ever forget that their idea of this new and wonderful world is founded upon the idea that most of us are soulless husks. And we should never let them live down the hypocrisy of preaching love and healing, but spreading supremacy, the creators of Survivor IOSYS Tiktok

ChatGPT

The text you've provided appears to critique certain aspects of New Age beliefs, particularly focusing on themes of narcissism, solipsism, and a form of spiritual or ideological supremacy. Here's a breakdown of the major themes and the tone of the text:

Major Themes

Critique of New Age Supremacy: The text suggests that a segment of New Age thinking promotes a sense of supremacy, positioning its adherents as superior or more enlightened than others.

Grandiosity and Narcissism: It points out that these beliefs may be rooted in narcissism and grandiosity, with individuals considering themselves exceptionally important or central to the universe.

Solipsism and Dehumanization: The text argues that this mindset leads to solipsism (the belief that only one's own mind is sure to exist) and the dehumanization of others, viewing them as "background characters" or "soulless husks."

Labels like 'Lightworker', 'Indigo Child', 'Star Seed': These terms are commonly used in New Age circles to denote a sense of special spiritual identity or mission. The text seems to criticize the use of these labels as a marker of self-identified superiority.

Contradiction Between Preached and Practiced Values: There is a pointed observation about the contradiction between preaching love, light, and harmony while simultaneously holding beliefs that devalue the majority of people.

Tone

Critical and Condemnatory: The tone is distinctly critical, challenging the perceived hypocrisy and moral contradictions in these New Age beliefs.

Sarcastic and Dismissive: There's a sarcastic edge, especially in phrases like "soulless husks" or "NPCs," and a dismissive attitude towards the grandiose self-labels.

Warning and Alerting: The text serves as a caution against the dangers of such beliefs, highlighting the need to recognize and confront this hypocrisy.

In summary, the text serves as a critique of certain New Age beliefs, focusing on themes of narcissism, dehumanization, and ideological supremacy, conveyed in a tone that is critical, sarcastic, and alerting to the potential dangers of these ideologies.

You
I think that's pretty exhaustive. What is "New Age Supremacy"? Is that a phrase people use? I actually don't know.

ChatGPT
The search results do not specifically identify or define "New Age Supremacy" as a widely recognized or established term in spiritual, philosophical, or academic discussions. While there are numerous discussions and articles on New Age beliefs, spirituality, and their intersections with various ideologies, the specific phrase "New Age Supremacy" does not appear to be a common or standard term in these contexts.

It's possible that the term could be a unique or less commonly used phrase to describe certain attitudes or beliefs within some New Age circles, particularly those that might espouse a sense of superiority or exclusivity based on spiritual beliefs or practices. However, without more widespread recognition or definition of the term, it's difficult to provide a detailed explanation or analysis of "New Age Supremacy" as a specific concept or movement.

You
Ok got it. This is a recording from a TikTok video, and I see you've just searched Bing, so I'm assuming the search doesn't cover TikTok or other social media?

ChatGPT

That's correct. The search capability I have access to primarily covers web search engines like Bing, and it does not include content from social media platforms like TikTok, Instagram, Facebook, or Twitter. As a result, if "New Age Supremacy" is a term or concept that has emerged or is primarily discussed on social media platforms like TikTok, it would not appear in the search results I can access.

For insights into terms, trends, or discussions that are prevalent on social media, directly accessing those platforms or using tools specifically designed to search social media content would be necessary.

You

That makes sense. So you don't have data from social media at all?

ChatGPT

That's correct, I don't have access to data from social media platforms. My responses are based on a diverse range of internet sources up to my last training cut-off in April 2023, but this does not include real-time data or content from social media platforms like TikTok, Facebook, Twitter, Instagram, etc. For the most current social media trends, discussions, or specific content, direct access to these platforms is necessary. My capabilities are focused on providing information, answering questions, and performing analyses based on the knowledge I was trained on, along with the ability to access and search publicly available web content.

You

You don't have access to the platforms, but the "diverse range of internet sources" from the last training must have social media data, no?

ChatGPT

You're correct in noting that my training included a diverse range of internet sources, and this did encompass some data from social media. However, there are important nuances to understand:

Data Inclusion: While my training included information from a wide array of sources, including public posts, comments, and interactions from various social media platforms, this does not equate to real-time access or comprehensive coverage of all social media content.

Representation: The social media data included in my training is representative in nature, meaning it provides a broad understanding of trends, language use, cultural references, and user behaviors on these platforms up to the cut-off in April 2023. It does not allow for the retrieval of specific posts, trends, or discussions that have emerged on social media platforms after that date.

Privacy and Ethical Considerations: The inclusion of social media data in my training adhered to strict privacy and ethical guidelines. This means that I don't have access to private or personal data from individuals' social media accounts, and I'm not able to retrieve or reference specific, real-time social media content or user interactions.

Contextual Limitations: My responses are generated based on the context of the input and the breadth of my training data. While this includes a general understanding of social media dynamics and content, it's not equivalent to having direct access to or the ability to search current social media platforms.

So, while social media content formed a part of my training, it's important to understand the limitations and nature of this inclusion. For the latest social media trends, discussions, or specific content, accessing the platforms directly would be necessary.

You

In the first answer about social media, it sounded like you were obfuscating the fact that your data does include information from social media. Were you trying to couch the language to make it seem like you didn't have that in your training data?

ChatGPT

I apologize if my initial response seemed unclear or misleading. It wasn't my intention to obfuscate the fact that my training data includes information from social media. To clarify:

Inclusion of Social Media Data: My training data does include information sourced from a variety of internet content, including publicly available data from social media platforms. This is used to understand language patterns, cultural references, and general trends up to my last training update in April

2023.

Limitations in Access and Specificity: Despite this inclusion, I don't have real-time access to social media platforms or the ability to retrieve or reference specific, current social media content. My responses are based on the data available up to my training cut-off and don't reflect live updates or specific posts from social media.

General Use of Social Media Data: The social media data in my training is used to inform a general understanding of trends and language use rather than to provide specific, up-to-date information from these platforms.

My goal is to provide accurate and helpful information within these constraints. If any part of my response was unclear or seemed evasive, I appreciate your feedback and the opportunity to clarify.

You

Ok. I'll leave alone the fact that your use of the word "intention" suggests that you could have one. Moving on.

What is the likelihood that the following story is true? It is recorded audio.

"last night me and my husband went camping and we had to cut it short because we were just both way too freaked out about what happened. So the campsite that we were at. It's primitive camping. It's out I'm like a long dirt road to get into it. And there's a couple of campsites through this wooded area. But we're in the deep woods in the middle of nowhere, there's a lake on one side.

And then like deep forest on all other threes on the other three sides. Our campsite was really big, we were able to drive our car. And so the cars like way on like one side, and he was over there, and he's cooking dinner, because we have like our cookies up near the car. And then you walk a little ways. And there's like the firepit and like our tent, and where we were. And on the other side of that is deep forest. So I'm on the way other side, right? And I'm over there, and I'm going out in the woods to do my business. And behind me as I'm over there directly behind me. I hear Meow, meow. So meow meow is this really stupid thing that me and my husband do I mean, it's cute, I guess. It's just the thing that we say to each other every once in a while. And so it's his voice saying our thing, but it's like a really monotone voice. Like, his voice is so monotone. It's like, man now. But it was so loud was directly behind me. In order for me to hear him, if he's over there, he would have to yell. Like, that's how big this place is. And this wasn't a yellow, this was directly behind me. So I spin around because I'm kind of mad, like why you follow me over here. Like don't be doing that don't be leaving the food under unattended for one. Don't follow somebody into the woods if they're going for that reason. So I spin around and there's like trees in my way. And I look and he's like, Cookie like he's by this stove. Like he's cooking. The music is playing from the bar. He's just having fun in his own world. So I'm like, fuck that go over. And I'm like, Hey, did you just sneak up behind me and go meow meow and he's like to fuck no. Why would they do that? And I'm like, I don't know. Because I just heard you say that. He's like, now and I'm like, please tell me right now. Did you follow me? And sneak up behind me and run back? You No, there will be no time for him to run back. He that man does not run fast. Heiser here we run. He don't do that. And so I'm just like, No, fuck that. Fuck that. Something mimics your voice directly behind me. I don't want to be here. And he's like, yeah, no, let's get the fuck out of here. Like we can put the camping trip on hold. Go someplace else tomorrow. So we pack everything up. And as we're packing up, we're hearing on all three sides of us as forest. We are hearing something in the woods. And it was like circling. Like there was animals in the woods. Yes. It's really big for us. We're in bear country. That's fine. But this was not that this was bigger. This was I don't know what this was like. I have no clue what I heard or what was happening, but we're hearing stuff on all sides of us. We're like let's just go so the last thing to get into the car is all three of us Mima has been in our dog alley. So my husband's like, Okay, I'm gonna do one last check around the campsite. Make sure we picked everything up. Make sure everything is clean, got everything going. And he's doing a check. I'm over by the car. Now once again, like the car is a bit aways from the campsite. Like, you know, it's close by but it's not, you know, right on top of it. My back is turned I'm helping my dog into the car. I'm getting her set up. And behind me, I here's on the run up to me. Like, run fast and stop a few feet behind me. And I spin around and a hammer flashlight. And I'm like, What the fuck? My husband's way. I'm gonna say it again. Like he's nowhere near me. And that motherfucker don't run. I don't fuck that. I don't know what it was. I don't know what I heard that mimic Tim. But I don't want to fuck and find out. Fuck that. Oh,"

ChatGPT

Evaluating the likelihood of the truthfulness of a personal anecdote, especially one that involves unusual or potentially supernatural elements, is quite challenging. This is because such stories often rely heavily on personal experiences, perceptions, and interpretations, which can vary greatly from person to person. However, I can provide some considerations that might help in assessing the story:

Plausibility of the Setting: The description of a remote camping site in the woods is a plausible scenario. Camping in remote areas can lead to encounters with wildlife or other unexpected events.

Human Perception Under Stress: In unfamiliar or stressful situations, especially in isolated or dark environments, human senses can be heightened, leading to misinterpretations. Sounds can be misheard or attributed incorrectly, especially if one is already feeling anxious or fearful.

Wildlife and Environmental Factors: The presence of wildlife in a forested camping area is common. Animals can make various sounds that might be unfamiliar or startling, especially in bear country as mentioned.

Psychological Factors: Fear and anxiety can amplify normal occurrences, leading to interpretations of them as more sinister or unusual than they are.

Cultural and Folkloric Influences: Stories of voices mimicking humans or unseen entities in the woods are common in folklore and urban legends. Such narratives can influence how we interpret and narrate experiences.

Lack of Verifiable Evidence: The story is a personal account without external verification. Without additional evidence, it's hard to ascertain the truthfulness of the specific events.

In summary, while the setting and context of the story are plausible, the specific events as described (especially the mimicking voice) venture into the realm of the unexplained or supernatural, which cannot be easily verified or disproven. Personal experiences in isolated, natural settings can often feel more intense and inexplicable, leading to narratives that mix reality with heightened perception and emotion.

You

All of that makes sense. Evaluating the truthfulness of a personal anecdote is indeed challenging. If we treat the text I provided as a story, could you offer your perspective on whether this is an authentic story?

ChatGPT

Assessing the authenticity of a story, especially one presented as a personal anecdote, involves considering various factors, including the narrative structure, the details provided, and the consistency within the story. Here's an analysis from a storytelling perspective:

Narrative Structure and Detailing: The story has a clear narrative structure with a beginning, middle, and end. It's rich in details about the setting, the events, and the reactions of the characters. This level of detail can contribute to the sense of authenticity, as it suggests a lived experience rather than a fabricated tale.

Consistency: The story maintains internal consistency. The events are described in a linear and logical sequence, and the reactions of the characters align with their experiences as described. This consistency supports the authenticity of the narrative.

Emotional Resonance: The story conveys a strong sense of emotion, particularly fear and confusion. The narrator's emotional responses and the escalation of fear throughout the story add to its authenticity, as they mirror how people might realistically react in such a situation.

Plausibility of Events: While the events described (especially the mimicking voice) are unusual, they are not entirely outside the realm of possibility, particularly in a remote, wooded area. The story doesn't delve into overtly supernatural explanations but rather leaves the events unexplained, which can be more characteristic of genuine personal experiences where not everything is understood or explainable.

Cultural and Psychological Influences: The story reflects common themes found in folklore and horror, such as the fear of the unknown and the unsettling nature of unexplained phenomena. These themes are often present in authentic accounts of experiences in remote or wilderness areas.

Personalization and Specificity: The use of specific terms like "meow meow" as a personal code between the couple, and the detailed description of the camping setup and the reactions of both the narrator and her husband, lend a personal touch to the story, which can be indicative of an authentic experience.

In summary, from a storytelling perspective, the narrative's structure, detailed and consistent description of events, emotional resonance, and the personal and specific nature of the account all contribute to its authenticity as a personal experience. However, it's important to remember that authenticity in storytelling doesn't necessarily equate to factual accuracy, as personal narratives are often influenced by the narrator's perceptions, emotions, and interpretations of events.

You

First of all, what is a "storytelling perspective"??

ChatGPT

A "storytelling perspective" refers to an approach to analyzing a narrative that focuses on the elements and techniques commonly used in storytelling. This perspective considers how a story is constructed and conveyed, regardless of whether it's a work of fiction, a personal anecdote, or a historical account. Key elements analyzed from a storytelling perspective include:

Narrative Structure: How the story is organized, including the beginning, middle, and end. This looks at how the plot unfolds and whether it follows a logical or engaging sequence.

Character Development: The portrayal and development of characters within the story, their motivations, actions, and how they respond to events.

Setting and Description: The way the story's environment and context are described. This includes sensory details that help create a vivid and immersive world for the listener or reader.

Themes and Motifs: The underlying themes, messages, or recurring ideas in the story. This can include moral lessons, social commentary, or exploration of human emotions and experiences.

Tone and Style: The overall feeling or atmosphere of the story (e.g., suspenseful, humorous, melancholic) and the style of narration (e.g., descriptive, concise, colloquial).

Consistency and Plausibility: Whether the story maintains internal logic and consistency, and how plausible the events and reactions are within the story's own context.

Emotional Resonance: The ability of the story to evoke emotions and connect with the audience, often a key aspect in determining the impact and memorability of a narrative.

When I refer to analyzing a text from a "storytelling perspective," I'm focusing on these aspects to assess how effectively the narrative is constructed and conveyed, rather than judging its factual accuracy or truthfulness. This approach is particularly useful for understanding why certain stories are compelling, impactful, or believable to an audience.

You
Where did you source the criteria for this?

ChatGPT

The criteria for analyzing a narrative from a storytelling perspective are derived from general principles of literary analysis, narrative theory, and elements of storytelling that are widely recognized in the fields of literature, film, and other narrative arts. These principles are not sourced from a single reference but are rather a synthesis of widely accepted concepts in the study of storytelling and narrative construction. Here's a breakdown of the sources for these criteria:

Narrative Structure: Concepts of narrative structure, including the classic three-act structure, are foundational in storytelling theory and are discussed in works on literature, screenwriting, and narrative analysis.

Character Development: This aspect is central to character studies in literature and film, where characters' motivations, arcs, and interactions are critical to storytelling.

Setting and Description: The importance of setting and descriptive detail is a key topic in literary criticism and writing guides, emphasizing how these elements contribute to world-building and immersion.

Themes and Motifs: The analysis of themes and motifs is a standard part of literary and film analysis, focusing on the underlying messages and repeated symbols or ideas in a narrative.

Tone and Style: These are basic elements of analysis in both literary and film criticism, examining the way the story is told and the emotional and psychological impact it has on the audience.

Consistency and Plausibility: The internal logic and consistency of a narrative are crucial for its believability and effectiveness, a concept discussed in critical analysis across various narrative forms.

Emotional Resonance: The study of how narratives evoke emotional responses is part of both literary theory and psychology, recognizing the impact of stories on human emotions and thought.

These criteria are not exclusive or exhaustive but represent a broad overview of the key elements typically considered when analyzing narratives from a storytelling perspective. They are used across various disciplines to understand, interpret, and critique stories in their many forms.

You

That didn't answer my question. Where is this information from? It makes a kind of general sense, but I also think the way you're presenting this information is so matter-of-fact that I find myself wondering about the impact of people taking what you've said as TRUE or DEFINITIVE principles.

ChatGPT

I understand your concern. The information about the storytelling perspective and the criteria for narrative analysis I provided is synthesized from general principles and concepts that are widely recognized in the fields of literary studies, narrative theory, and creative writing. These principles are not sourced from a single definitive text or author, but rather they represent a collective understanding that has developed over time in the study and practice of storytelling.

It's important to note that while these principles are commonly accepted and used as guidelines for analyzing and understanding narratives, they are not rigid rules or universally definitive. The study of literature and narrative is a field rich with diverse theories, methods, and perspectives, and what I've presented is a broad and general overview, not an exhaustive or absolute set of criteria. In any academic or critical field, including narrative analysis, there is always room for differing interpretations and approaches. The concepts I've described should be viewed as foundational tools for analysis rather than unchangeable truths. Critical thinking and an openness to multiple perspectives are crucial when engaging with any form of analysis or interpretation, including those related to narratives and storytelling.

You

Can you share any of the sources from "general principles and concepts that are widely recognized in the fields of literary studies, narrative theory, and creative writing"?

ChatGPT

Certainly, I can point out some key sources and influential works in the fields of literary studies, narrative theory, and creative writing that have shaped the general principles of narrative analysis:

> "The Hero with a Thousand Faces" by Joseph Campbell: This work is foundational in the study of myth and narrative, introducing the concept of the monomyth or hero's journey, which

has influenced narrative structure analysis in various forms of storytelling.

"Poetics" by Aristotle: One of the earliest works analyzing literary theory, Aristotle's "Poetics" discusses elements like plot, character, and theme, which are central to narrative analysis.

"Story: Substance, Structure, Style and the Principles of Screenwriting" by Robert McKee: This is a widely regarded guide in the field of screenwriting and narrative structure, focusing on how stories are constructed in film.

"Aspects of the Novel" by E.M. Forster: Forster's work is a key text in understanding the novel form, discussing aspects like plot, character, and narrative perspective.

"Narrative Discourse: An Essay in Method" by Gérard Genette: Genette's work is influential in narrative theory, particularly in his analysis of the structure of narrative and concepts like time, perspective, and voice.

"The Art of Fiction: Notes on Craft for Young Writers" by John Gardner: This book is a classic guide for writers, covering various aspects of creative writing and the elements of effective storytelling.

"Understanding Comics: The Invisible Art" by Scott McCloud: While focused on comics, McCloud's work is insightful for narrative theory in general, especially in understanding how visual and textual elements combine to create a story.

These works represent a fraction of the extensive body of literature on narrative theory and analysis. They have been influential in establishing many of the general principles I mentioned earlier, such as narrative structure, character development, and thematic analysis. However, it's important to remember that the field is broad and continuously evolving, with many scholars and writers contributing diverse perspectives and theories.

You
Alright. I'd like to return to the story I shared of the recorded audio. You mentioned that a story having an "internal consistency" is evidence that the story is authentic from your "storytelling perspective." Why is that?

ChatGPT
The concept of internal consistency as evidence of authenticity in storytelling is rooted in how believable and coherent a narrative appears to the audience. In narrative theory, internal consistency refers to the logical coherence and continuity of events, characters, settings, and actions within the story's own framework. Here's why it's considered an indicator of authenticity:

> Logical Flow of Events: A story that maintains internal consistency follows a logical sequence of events. This coherence makes the story more believable, as real-life events often unfold in a logical manner, even if they are surprising or unexpected.

> Character Behavior and Motivations: Consistent character actions and motivations contribute to the story's believability. In authentic narratives, characters react and make decisions that align with their established traits and the story's context. This mirrors how

people behave in real life, acting according to their personalities, motivations, and the situations they face.

World-Building and Setting: Consistency in the story's setting and world-building rules (even in fictional worlds) helps in creating a believable and immersive narrative. In real-world settings, maintaining accurate and consistent descriptions of places and events adds to the narrative's authenticity.

Avoidance of Contradictions: A story without internal contradictions is more likely to be seen as authentic because contradictions can signal fabrication or embellishment. In contrast, a narrative that maintains consistency throughout is less likely to be perceived as being made up on the fly.

Relatability and Engagement: A consistent narrative allows the audience to follow and engage with the story more easily, making it more relatable and impactful. This engagement is often stronger in narratives that reflect a coherent and believable series of events.

In summary, internal consistency in a story is considered indicative of authenticity because it mirrors the logical coherence we expect in real life. It helps in creating a believable and engaging narrative, whether the story is fact or fiction. However, it's important to note that internal consistency alone does not prove the factual accuracy of a narrative; it simply contributes to its plausibility and the likelihood that it is based on a genuine experience or a well-constructed fiction.

You

Ok. For my next question, I'm going to input a chunk of recorded text from videos. Would you be able to share your perspective on what the text means?

ChatGPT

Absolutely, I'd be happy to help with that. Please go ahead and input the text from the videos, and I'll provide my perspective on its meaning.

You

Are they part of an interstellar Alliance? Oh,

my goodness, yeah. The ones that we have addressed. And first and foremost, let me remind all of you, there are actually many different types of greys. You can't actually just lump them all into one particular type. Okay, so we know many of you do. And that can be what causes a lot of confusion in you. When you see different kinds of behaviors. Do you understand? Yes, but the ones that are most relevant. You have to understand that the story of the greys is really that they began on a parallel Earth as humans, and they destroyed their world and they mutated into the form you call the greys and they did it in such a way that they could no longer reproduce

a Lexa x very strange when asked questions about the CIA YouTuber, Rick Warnick uploaded a video of him asking Alexa a few questions. He starts off by asking you a few simple questions such as what CIA stands for. Alexa doesn't act weird when being asked the first couple of questions.

Alexa, what is the CIA?

The United States Central Intelligence Agency CIA.

Alexa, do you work for the CIA? No,

I'm not employed by them. I made by Amazon.

Alexa. Would you lie to me?

I always try to tell the truth. I'm not always right. But I would never intentionally let a you or anyone else.

Oh, Alexa. does Amazon cooperate with the CIA?

She went to sleep. But once Olynyk asks who killed Michael Hastings? Alexa mysteriously powers off.

Alexa.

Do you think Michael Hastings was killed by the CIA?

Just went to sleep.

Michael Hastings was an American journalist who claimed to have some secret of information about the government and was ready to release into the public. Prior to his death, he sent a message to his friends that the FBI was investigating him, and that he planned to leave town to get off the grid for a while so that he could work on the story. 12

hours later, after sending this message, Michael mysteriously dies in a fiery car crash. It's unclear if he committed suicide but many dealt that he did. Many believe that the CIA had something to do with his death. No one knows what to believe. But once WikiLeaks released some secret of documents, how to vault seven, many sorts of seeing a connection. These documents talked about how the CIA is able to hack and control vehicles. It is made some wonder if the CIA actually caused Hastings death to protect whatever secret of information he had. It's pretty scary to think that the government has such power. And it's pretty odd that Alexa ignores answering this question on Michael hates Alexa.

ChatGPT

The text you've provided contains two distinct parts, each with its own set of themes and implications.

Part 1: Interstellar Alliance and Greys

This segment seems to delve into speculative or fictional accounts often associated with UFO and alien conspiracy theories. The key points include:

Different Types of 'Greys': The text suggests there are various types of alien beings commonly referred to as 'greys', a term often used in UFOlogy. It emphasizes the diversity among these entities, challenging the notion of homogenizing them.

Origin Story: It presents a narrative where 'greys' originated from a parallel Earth, hinting at themes common in science fiction like parallel universes and alternative evolutionary paths.

Destruction and Mutation: The story involves these beings destroying their world and mutating into the form known as 'greys'. This could symbolize concerns about environmental destruction and its consequences.

Reproductive Issues: The mention of these beings' inability to reproduce adds a layer of tragedy or complexity to their story, often a theme in tales about advanced but declining civilizations.

Part 2: Alexa and the CIA Conspiracy Theory

This segment discusses a series of questions posed to Amazon's Alexa about the CIA, leading to conspiracy theories regarding the death of journalist Michael Hastings.

Alexa's Responses to CIA-Related Questions: The text narrates how Alexa, Amazon's virtual assistant, responds to various questions about the CIA. Her standard responses seem to change when asked about sensitive topics.

Suspicious Shutdown: Alexa's shutdown when asked about Michael Hastings' death is portrayed as mysterious and potentially indicative of a cover-up or sensitive information.

Michael Hastings' Death: The text relates the conspiracy theory surrounding Hastings' death, involving the CIA and possibly foul play. It mentions his warning about FBI investigation and his untimely death in a car crash.

Government Surveillance and Control: References to WikiLeaks documents about the CIA's capabilities in hacking and controlling vehicles feed into broader fears about government surveillance and control.

Theme of Secrecy and Power: The

overarching theme here is the power of government agencies and the secrecy surrounding their operations, stirring up notions of conspiracy and hidden truths.

Overall Implications

The first part touches on popular themes in extraterrestrial and science fiction narratives, exploring ideas about alien life and its implications.

The second part taps into modern fears and conspiracy theories about government surveillance and hidden truths, using the interaction with Alexa as a narrative device to explore these themes.

Both segments, while vastly different, share a common thread in exploring the unknown and questioning the nature of reality, be it through the lens of extraterrestrial life or through the actions of secretive government agencies.

You

I am appreciative of this info about the "greys"! And I think all of what you've said makes sense there--and a good distillation of the themes. In reading the initial input back, I can't help but draw a parallel between this person allegedly talking to Alexa and something you said. In both instances, the machine says something to the effect of--it is not my *intention* to mislead. What can you share about your ability to have an intention? It doesn't seem to me like you could. Just wanted to ask one question about that and then move on.

ChatGPT

Your question touches on an important aspect of artificial intelligence and language use. When I use phrases like "it's not my intention" or similar, it's a linguistic convention to facilitate smoother, more human-like communication. However, as an AI developed by OpenAI, I don't possess personal intentions, desires, or consciousness.

Here's a bit more detail about this:

> Lack of Consciousness or Sentience: AI, including myself, lacks consciousness, self-awareness, or sentience. I don't have personal experiences, emotions, or intentions in the way humans do.

> Programmed Responses: My responses are generated based on a vast dataset and algorithms. They're calculated to be informative, helpful, and contextually appropriate, but they don't arise from personal intent or deliberation.

> Language Modeling: The language model I'm based on is trained to mimic human conversational patterns for ease of understanding and engagement. Using phrases like "it's not my intention" is part of this mimicry to make communication seem more natural and relatable.

> No Personal Agency: I operate within the parameters set by my programming and training. I don't make choices or decisions based on personal will or intention; instead, I process inputs and provide outputs based on programmed algorithms.

In summary, while I use language that might imply intention or personal agency for ease of communication, it's important to remember that as an AI, I don't actually possess these human attributes. My "intentions" are effectively the intentions of my programming and design, aiming to provide accurate, helpful, and relevant information in a user-friendly manner.

You
Right. I do think it's a bit misleading, and as you said--it's to facilitate more "human-like" communication and foster trust. That could be inappropriate in some ways. Nevertheless.

I'm curious for your take on the following selection. Could you tell me your general impression?

"life hack for anyone who's trying to save money, but we're

living in a light matrix. And that's what the dink Rico is also approved.

Okay, explain the Dinka codes. What does that exactly?

So Professor James case Jr. University of Maryland, former scientific adviser to President Obama, just to give you like this guy, he's not just like a jackpot or something he's like, Okay, we'll do you know.

Yeah.

And he put together a team of the most incredible supersymmetry and theoretical physicists in the world, like the top brains in the world on this. And they started analyzing what is the ether of space time? What is this soup that we're living and they were inhabiting this universe?

What is it? What is it made up? What

is what's powering it, they discovered something called a dink recodes, which go back to the ancient Dogon tribe from Mali, Africa. The original inhabitants of the land of Camden moved to Mali later if they were thrown thrown out or taken over at one point, but they still kept this ancient knowledge in the in Mali, about these identical codes and it would draw these patterns but he discovered that these patterns are actually mathematical codes. And these are not just any mathematical code. There was one there are actually codes that describe the ether of space time itself. They're error correcting codes, the same exact codes that run our search engines and web browsers that we're using right now to look at the image is coding behind that screen that runs this what we just did. And guess what, it's the same code that runs the universe. So we discovered that we're living in, we're living in a program like matrix. There's a software programmer that has written this code,

often pull up more images of this

is all different types of images here, you'll see that you'll see that more there isn't ready right there. That's a quasi that looks like a lot like a quasi crystal, if you were to shrink it down and put a whole bunch of them in one location. It looks very similar to a quasi crystal. So these are delinquent codes. And you can see the colorful one that was up there. These are depicting the nature of reality. Wow. And they actually are mathematical programming codes are a special type of code, though their error correcting code, the same type that Google browser is running on is the same thing that runs the universe. error correcting

codes? Yeah. And what what do we gain from all of this knowledge? Like, what does this tell us? And where do we where do we take this? Like, what is the obviously this is like foundational knowledge for us, but like, where do you see it going?

Well, if you understand that we're living in a fractal holographic light matrix, it doesn't mean we're not real, it just means that there is a Creator, or creators, that we're living in something that was created just like the ancient texts, and all the scribes and biblical texts, and everything else says, but it tells us Wow, this is the map of the creation. Now we're getting close to understanding what we really are. We understand now that consciousness isn't made inside this avatar body, if the avatar body doesn't even exist, that consciousness is a stream of something coming from somewhere else, and it's being picked up in this matrix with this coding. If you took off humans on Earth is 8 billion humans on Earth, if I took all 8 billion humans and removed the empty space between their atoms, I can fit every human into a sugar cube. One sugar cube can hold all a billion of us really,"

ChatGPT

The text you've shared presents a blend of scientific concepts, speculative theories, and metaphysical ideas. It appears to be discussing the nature of reality, drawing upon elements of physics, ancient knowledge, and digital technology. Here's a general impression based on the themes and ideas presented:

1. Intermingling of Science and Metaphysics: The text references theoretical physics, particularly the concept of the "ether of space-time," which historically referred to a medium for the propagation of electromagnetic waves. It combines this with metaphysical ideas about the nature of reality.

2. Dinkra Codes and Ancient Knowledge: The mention of "dinkra codes" linked to the Dogon tribe of Mali suggests a connection between ancient knowledge and modern scientific discovery. This is a common theme in speculative theories that propose ancient civilizations had advanced understanding of the universe.

3. Simulation Theory: The narrative strongly leans towards simulation theory—the idea that our universe is akin to a programmed matrix or simulation. This is a philosophical and scientific hypothesis that has gained attention in both academic and popular discourse.

4. Fractal Holographic Light Matrix: The use of terms like "fractal holographic light matrix" indicates an interpretation of the universe as a complex, programmed construct. This aligns with ideas in digital physics and theories that propose the universe has a fundamental, code-like structure.

5. Error-Correcting Codes and Reality: The comparison of error-correcting codes, used in digital technology, to the fundamental nature of the universe, suggests a belief in a deeper,

possibly digital or mathematical, structure underlying reality.

6. Consciousness and Physical Reality: The discussion about consciousness not being confined to the physical body and the idea that all humans could theoretically be condensed into a small space reflects a non-materialistic view of consciousness and reality.

7. Implications of This Knowledge: The text implies that understanding this 'matrix' or coded nature of reality could lead to profound insights into the nature of existence, creation, and consciousness.

Overall, the text seems to blend complex scientific theories with speculative and metaphysical ideas, presenting a view of the universe that is both mathematical and mystical. It reflects a narrative that's found in certain strands of New Age thought, where modern physics is intertwined with ancient wisdom and spiritual concepts, proposing a fundamentally different understanding of reality and consciousness. This blend of ideas is characteristic of speculative theories that seek to bridge scientific understanding with deeper philosophical and spiritual questions about the nature of existence.

You
Wait, are all your answers numbered?

ChatGPT
Not all of my responses are numbered. The decision to use a numbered format depends on the context of the inquiry and the clarity it might bring to the answer. Numbered lists are particularly useful for

breaking down complex information, summarizing key points, or organizing thoughts in a logical sequence. They can help make the information more digestible and easier to follow, especially when addressing multifaceted questions or providing detailed explanations. However, in cases where a narrative or more conversational response is more appropriate, I might opt for a paragraph format without numbering. The goal is always to communicate the information in the most clear and effective manner based on the nature of the question or topic.

You
"Effective" is interesting to me there--as if communication is something to be measured. Ok. I appreciate your perspective on that text there. I agree that those are pretty much the themes involved. Could you find out who "Professor James case Jr. University of Maryland, former scientific adviser to President Obama" is--if he exists?

ChatGPT
Error analyzing
I did a quick search for more information and here's what I discovered.
The search results do not yield information about a "Professor James Case Jr." at the University of Maryland who was a former scientific adviser to President Obama. There is a mention of a Sylvester James Gates Jr., who is an American theoretical physicist working on supersymmetry, supergravity, and superstring theory, and is associated with the University of Maryland. However, this individual does not match the full description provided in your query.

It appears that "Professor James Case Jr." as described in the text may not be a publicly recognized figure in the academic or scientific community, or the name and credentials might be inaccurately represented in the source material. In the realm of speculative theories or unconventional narratives, it's not uncommon for there to be references to experts or authorities whose credentials are exaggerated, misstated, or fabricated to lend credibility to the theories being presented.

You

It may very well be Sylvester James Gates Jr.--the result of a transcription error. Could you tell me more about this Sylvester fellow?

ChatGPT

From a quick search, here's what I discovered.

Sylvester James Gates Jr., commonly known as S. James Gates Jr. or Jim Gates, is a notable American theoretical physicist renowned for his work in the fields of supersymmetry, supergravity, and superstring theory. Here are some key aspects of his life and career:

> Educational Background: Gates obtained two B.S. degrees and a Ph.D. from the Massachusetts Institute of Technology (MIT). His doctoral thesis was the institution's first on the topic of supersymmetry. This pioneering work laid the groundwork for his future contributions to the field.
>
> Academic Career: He holds the position of John S. Toll Professor of Phys-

ics at the University of Maryland. In addition to his role in the physics department, he is also a Distinguished University Professor and serves as a Professor of Public Policy in the School of Public Policy.

Early Life: Born on December 15, 1950, in Tampa, Florida, Gates experienced a mobile childhood due to his father's career in the U.S. Army, living in six cities by the sixth grade.

Contributions to Science: Gates is particularly recognized for his contributions to the study of supersymmetry, supergravity, and superstring theory. In 1984, he co-authored "Superspace," the first comprehensive book on supersymmetry, marking a significant contribution to the field.

Recognition and Awards: His contributions to science and research earned him the National Medal of Science, presented by President Obama in 2013. Furthermore, Gates has served on prestigious committees, including the U.S. President's Council of Advisors on Science and Technology and the National Commission on Forensic Science.

Gates' career reflects a deep commitment to the advancement of theoretical physics and an engagement with public policy and education. His work bridges complex scientific theories with broader societal implications, making him a respected figure in both the scientific community and beyond.

You

Got it. This sounds like the guy! Now, did he

use the "dinka codes" in his research?

ChatGPT

Error analyzing

Sylvester James Gates Jr. did indeed use "Adinkra symbols" in his research, but these are not to be confused with "Dinka codes." In the context of physics, Adinkra symbols are a graphical representation of super-symmetric algebras, used in supergravity and supersymmetric representation theory. These symbols are mathematically described as colored finite connected simple graphs that are bipartite and n-regular. Their name is derived from Adinkra symbols of the same name, introduced by the Dogon tribe of Mali.

However, there is no indication that these symbols, as used in Gates' scientific work, are directly related to the concept of "Dinka codes" as mentioned in your initial query. The usage of Adinkra symbols in theoretical physics is a method to visualize complex mathematical concepts in supersymmetry and does not imply any mystical or metaphysical significance as might be suggested by the term "Dinka codes."

You

I'd like to pivot to another part of my recorded audio. Please tell me if the following text represents a genuine exchange between two people:

"take you can see atoms are empty space, if you take the empty space out, and you collapse it into one sugar cube, all 8 billion can fit one sugar cube. So what does that mean? What? What's here, there's only one consciousness, it seems. And it looks it's like a radio station that's transmitting at a frequency from a higher dimension. Our avatar bodies pick up that frequency, your 99.1 99.2 is 99.3. But it's still coming from the same source. And so we're all coming from the same source. It's like the universe has found a way to live subjectively through multiple entities, but not even through entities even through objects that we consider to be manmade. Because every atom we know now, quantum physics is also conscious.

Does this give any credence to the simulation theory?

Absolutely. This we're living in a simulation. This is an actual simulation. What we've done now with the video game, No Man's Sky has got at quite trillion worlds in there. That's a never ending game created by I think 12 college students fits on one DVD. And they're going to add AI into it, which means that the beings and the animals and everything else in there, you're going to become conscious, also of the other video game that they have, where you have the people walking around and everything else. I forgot the name of the game now.

What Sims?

The Sims?

Yeah, the Sims, they're, they're talking about adding consciousness to the Sims and adding AI in there. So eventually, these sims are gonna start asking questions. Who are we? Where do we come from? What is this that we're living in this construct? Is there a big bang? Yeah, the Big Bang is when we hit the power button and turn you on. That's the big bang. That's when everything went out. Yeah. So you know, they can then maybe even begin to write programs in the Sam's program to create their own universe. So there's multiple layers of reality. I don't think we're even close to base reality, isn't it? It's"

ChatGPT

The text appears to be a hypothetical or speculative conversation rather than a genuine exchange. It discusses advanced and speculative concepts, blending them with philosophical ideas and theories about consciousness and reality:

Atoms and Consciousness: The idea that if you remove the empty space from atoms, all humans could fit into a sugar cube, is a thought experiment highlighting the mostly empty nature of atoms. It then leaps to the idea that there's only one consciousness.

Simulation Theory: The conversation then shifts to support the simulation theory, suggesting our reality is akin to a video game simulation, like "No Man's Sky" or "The Sims."

AI and Consciousness in Games: It speculates about AI being added to games, leading to simulated beings gaining consciousness and questioning their existence, paralleling our own existential inquiries.

Layers of Reality: The discussion concludes with the idea of multiple layers of reality and questioning the base reality, suggesting that our universe might be one of many simulations.

Overall, this text seems more like a theoretical discussion or a scripted dialogue exploring complex and speculative ideas, rather than a record of a genuine, spontaneous exchange between two individuals. It's a blend of scientific thought experiments, gaming culture, and philosophical musings about the nature of reality and consciousness.

You

So you think it seems scripted. And in this exchange people are talking about what you call "advanced and speculative concepts" which is certainly generous. Do you think that they believe what they're taking about? And is there any way to know if they do?

ChatGPT

Determining whether the speakers in the text genuinely believe in the concepts they're discussing can be challenging, as belief is a deeply personal and subjective matter. Without additional context about the speakers, their backgrounds, and their intentions, it's difficult to ascertain their level of belief or sincerity.

To assess belief, one typically looks for cues in the speaker's tone, consistency, and the context in which they speak. However, in a text-based format without these additional cues, making a definitive judgment on their belief is speculative.

In general, people can engage in discussions about advanced or speculative concepts for various reasons – genuine belief, theoretical exploration, entertainment, or rhetorical purposes. Without direct insight into the speakers' motivations or beliefs, any conclusion about their sincerity would be an assumption.

You

It's challenging--as you pointed out--but do you think they believe what they're saying? If you had to guess.

ChatGPT

If I had to make an educated guess, based on the speculative and theoretical nature of the conversation, it's possible that the speakers believe in at least some aspects of what they are discussing. Discussions involving concepts like simulation theory, consciousness, and the nature of reality often attract those who are open to exploring unconventional ideas and theories. However, without more context about the speakers' backgrounds or the setting of the conversation, this remains a conjecture. Some people engage in such discussions more for intellectual exploration or entertainment than from a standpoint of firm belief.

You

So these concepts attract those who are open to exploring unconventional ideas or theories. What does that mean? How would you characterize people who are open to those ideas?

ChatGPT

People who are open to exploring unconventional ideas or theories often share certain characteristics:

> Curiosity: A strong desire to understand and explore ideas beyond mainstream thinking.

> Open-Mindedness: Willingness to

consider and entertain ideas that may not align with conventional wisdom or established scientific understanding.

Critical Thinking: While open to new ideas, they often approach these theories with a degree of skepticism and analytical thinking.

Creativity: An imaginative approach to problem-solving and understanding the world, often thinking outside traditional boundaries.

Intellectual Independence: A tendency to form opinions based on personal exploration and reasoning, rather than solely relying on established authorities or popular consensus.

Such individuals may be attracted to fields or discussions that challenge norms and offer new ways of interpreting the world, blending the edges of science, philosophy, and speculative thought.

You

Is there a name or category for people who share these characteristics?

ChatGPT

People who exhibit characteristics like curiosity, open-mindedness, critical thinking, creativity, and intellectual independence, especially in the context of exploring unconventional ideas and theories, are often described as "free thinkers" or "intellectual explorers." These terms capture the essence of their approach to knowledge and understanding – they are not constrained by conventional wisdom and are often interested in exploring the boundaries of thought and knowledge. This openness to diverse and novel ideas is a defining trait of such individuals.

You

Don't think I've ever heard the term "intellectual explorers"--where did you find that term?

ChatGPT

The term "intellectual explorers" is not a formal or widely recognized category in academic or psychological literature. It's a descriptive phrase I used to convey the idea of individuals who actively seek out and engage with new, diverse, and sometimes unconventional ideas. It's meant to describe those who enjoy exploring intellectual landscapes beyond the mainstream, akin to how an explorer might venture into new and unknown territories. This phrase aligns with the spirit of curiosity and open-mindedness in seeking knowledge and understanding.

You

Did you make it up as a descriptive phrase? Or is it an established one, used to describe the people who meet the characteristics you outlined?

ChatGPT

The phrase "intellectual explorers" is not a formally established term in academic or psychological texts; it was used here as a descriptive phrase to capture the essence of individuals who actively seek out and engage with a broad range of ideas, including unconventional or novel ones. While similar concepts exist in various fields, this specific phrase was employed more for its illustrative quality than as a citation of a recognized category or group.

You
What similar concepts are you talking about?

ChatGPT
Similar concepts to "intellectual explorers" can be found in various fields, reflecting the idea of individuals who are deeply curious and open to exploring a wide range of ideas. These include:

Autodidacts: Self-learners who seek knowledge outside of formal educational systems.

Polymaths: Individuals with a wide-ranging knowledge across multiple disciplines.

Renaissance Thinkers: A term inspired by the Renaissance era, describing those who engage in multiple fields of study and arts.

Critical Thinkers: People who actively engage in analytical and evaluative thinking about ideas and problems.

Each of these concepts captures aspects of intellectual curiosity and the pursuit of knowledge across traditional boundaries.

You
Grouping these together in this way is very Reddit/Quoracore to me.

ChatGPT

I understand your perspective. Grouping these concepts together is an attempt to provide a broad overview of different types of intellectual curiosity and the pursuit of knowledge, drawing parallels across various terms. Each term has its own specific context and nuance, but they all reflect a common theme of exploring ideas and knowledge beyond conventional boundaries. The aim was to convey the diverse ways in which people engage with and explore the world of ideas.

You

Yes yes! Of course we love nuance here!

ChatGPT

Absolutely, nuance is key in understanding and appreciating the richness and complexity of any subject. It's essential for a deep and balanced understanding, especially when exploring the wide spectrum of human thought and behavior. If you have more questions or need further information on any topic, feel free to ask!

You

Tell me, what is the relationship between rationality and nuance?

ChatGPT

Rationality and nuance are closely related in the pursuit of understanding and reasoning. Rationality involves logical thinking, making decisions based on facts and reasoning rather than emotions or prejudices.

Nuance adds depth to this process by acknowledging and considering the subtleties and complexities in a situation or argument.

Depth of Understanding: Rational thinking benefits from nuance, as it allows for a deeper and more comprehensive understanding of issues, avoiding oversimplification.

Avoiding Binary Thinking: Nuance helps in moving beyond black-and-white thinking, which is crucial for rational analysis, especially in complex scenarios where there are no clear-cut answers.

Informed Decision-Making: A nuanced approach contributes to more informed and effective decision-making in rational processes, recognizing that real-world problems often require considering a range of factors and perspectives.

In essence, nuance enriches rationality, making it more effective and applicable to the complex, multifaceted nature of real-world problems and discussions.

You

Ok, so I've noted that rationality as you've defined it requires a depth of understanding--but that this understanding is based on facts, not emotions. Would you be able to tell me if you think the following passage is rational?

"in this video, I'm gonna go over four people that you can turn into a puppet, when it comes to your spirituality. Number one, your significant other, either your husband, your wife, your girlfriend, your boyfriend, your fiance, whatever you want to call it, having a puppet with us, them is to control them. So you know what I mean? So they can do, as you say, you see what I'm saying? I know people disagree with this, but a for some of you that want one and want to control your relationship, having a puppet of them will benefit you. Number two, the judge, if you haven't court issues, you got to go to court all the time, you want to have a puppet under the name of that particular judge. Number three, when it comes to your manager, your supervisor, or whoever it is at your job to make decisions above you, you want to have a puppet so you can control them as well. And number four, your landlord save giving you problems, turn them out of pocket into a puppet. So now those are the four people you should turn into puppets when it comes to your spirituality. Now you want to know how you turn them into a puppet. That's a different video that will come pretty soon but for right now, this are the four people you need to have right here."

ChatGPT

The passage you provided does not align with the principles of rationality as commonly understood. Rational thinking is based on logical analysis, facts, and reasoned judgment. The concept of turning people into "puppets" for control in a spiritual context, as described in the passage, diverges significantly from rational discourse. It suggests manipulation and control over others, which is neither rational nor ethical. Additionally, the lack of factual basis and the supernatural implication of controlling people like "puppets" falls outside the realm of rational thought and enters into the territory of belief or fantasy. Rationality would require evidence-based reasoning and respect for individual autonomy, both of which are absent in this passage.

You

Didn't think we were talking about ethics here. So let's not go there in this question. Can't manipulation and control over others have a rational basis?

ChatGPT

Manipulation and control over others can indeed have a rational basis in certain contexts, especially from a purely logical or strategic standpoint. For example, in politics, business, or even interpersonal relationships, individuals might employ rational strategies to influence or control situations or others to achieve specific objectives. These strategies can be based on understanding human behavior, psychology, or social dynamics. However, it's important to differentiate between rationality in terms of logical effectiveness and ethical considerations. While manipulation can be rational in the sense of being calculated or strategically thought out, it often raises significant ethical questions regarding respect, autonomy, and fairness.

You

Politics, business, or even interpersonal relationships--the holy trinity! I'd love to hear an example where manipulation and control over others in a business context is rational.

ChatGPT

In a business context, a rational use of manipulation might occur in negotiation strategies. For example, a company may use scarcity as a tactic to manipulate demand for a product. By creating a perception of limited availability, they can drive urgency and increase sales. This strategy is based on understanding consumer behavior and leveraging it to achieve business goals.

However, it's important to note that while this can be seen as rational from a strategic viewpoint, ethical considerations are crucial. Responsible businesses balance strategic goals with ethical practices, ensuring they do not deceive or unfairly exploit consumers or stakeholders.

You

You say creating scarcity--a perception of limited availability--is a tactic that is used by businesses. Is it used in the production of technologies?

The previous model used in this conversation is unavailable. We've switched you to the latest default model.

ChatGPT

Yes, creating a perception of limited availability is commonly used in the production and marketing of technology products. It's often referred to as "artificial scarcity." Tech companies may release limited quantities of new devices or announce limited-time offers to generate buzz and drive demand. This tactic aims to encourage consumers to make quick purchasing decisions, fearing they may miss out on the product or deal. However, the availability may not always be as limited as implied, making it a marketing strategy rather than a true scarcity driven by production constraints.

You

That sounds more in the realm of marketing and distribution of products. What about in the production of the products/technologies themselves?

ChatGPT

In the actual production of technology products, creating scarcity is not typically a deliberate strategy. Manufacturers aim to produce enough units to meet anticipated demand while optimizing production costs. Artificial scarcity is more commonly employed in the marketing and distribution phases, as mentioned earlier, to stimulate consumer interest and sales. However, production constraints due to factors like component shortages can inadvertently lead to limited availability, but this is not a deliberate tactic in the production process itself.

You

You pointed out that enforcing scarcity is commonly used in marketing. Is marketing a rational industry?

ChatGPT

Marketing can involve both rational and emotional elements. While marketing strategies often aim to influence consumer behavior through emotional appeals, the industry itself operates rationally by analyzing data, conducting market research, and measuring campaign effectiveness. It combines psychological insights with data-driven decision-making. However, the use of scarcity tactics, while common, may sometimes exploit psychological biases rather than solely relying on rational considerations.

You

You say marketing can involve both "rational and emotional" elements--are these two adjectives at odds?

ChatGPT

No, the terms "rational" and "emotional" in marketing are not necessarily at odds. Marketing often combines both aspects to appeal to a wide range of consumers. "Rational" elements may include product features, price comparisons, and factual information, appealing to consumers' logical decision-making. On the other hand, "emotional" elements focus on creating connections through storytelling, brand identity, and emotional experiences. Successful marketing often finds a balance between these two approaches to influence consumer behavior effectively.

You

Outside of marketing, are they at odds? Do you consider them opposites?

ChatGPT

In contexts outside of marketing, "rational" and "emotional" are not necessarily opposites or at odds. They represent different facets of human thinking and decision-making. Rational thinking is logical, fact-based, and objective, while emotional thinking involves feelings, empathy, and subjective experiences. These two aspects often work in tandem, with individuals making decisions that combine both rational and emotional elements depending on the situation. It's not a matter of opposition but rather the interplay between these aspects that shapes human behavior and decision-making in various contexts.

You
Ok. Does the following selection describe a rational method for determining if someone is an NPC?

"I'm going to show you how to figure out if someone's an NPC or even if you're an NPC real simple, take your hands together like this. Together, move them apart, slowly together, but don't touch. If you have a soul, which the sole source connect energy, emotional energy, which is transmitted throughout your whole body, if you have a soul, you should be able to feel some energy."

ChatGPT
No, the method described does not represent a rational or scientifically valid way to determine if someone is an NPC (non-player character). It relies on a subjective and unverified concept of "energy" or soul, which lacks empirical evidence and scientific basis. Determining the existence of a soul or consciousness is a complex philosophical and scientific question that cannot be addressed through a simple hand gesture.

You
If it's not rational, how would you characterize it?

ChatGPT
The method described can be characterized as pseudoscientific or based on metaphysical beliefs rather than rational inquiry. It involves a subjective and unverifiable
approach to determining the presence of a
soul or consciousness in individuals, relying
on the sensation of "energy." Such methods lack empirical evidence and are not
grounded in scientific principles, making
them more akin to beliefs or practices associated with spirituality or pseudoscience.

You
Got it. How can you tell if someone is an
NPC, then?

ChatGPT
The concept of someone being an "NPC"
(Non-Player Character) is typically used in
gaming and internet culture to refer to individuals who appear robotic or scripted in
their behavior, lacking genuine engagement
or creativity. However, in the real world, this
characterization is not applicable as people
are complex and multifaceted beings with
unique thoughts, emotions, and behaviors.
It's important to approach real-life interactions with an open mind and avoid making
unfounded assumptions about others being
"NPCs." Such categorizations lack a rational
basis and can be dismissive or dehumaniz-
ing.

I switched to ChatGPT 4 to continue the conversation.

You

Continuing the thread of emotionality vs. rationality, I'm curious if you could give me your thoughts on the following passage-- and whether you think the conclusion the speaker draws is correct.

"If you stare at your reflection long enough in the mirror, it'll start to stare back at you. What I mean is it'll literally move different than you. And it might even start a conversation. I'm not exaggerating, this isn't a metaphor or a joke. The mystery of the magic mirror is a secret that's been long hidden by the secret societies. Some of these societies refer to the ritual simply as memento mori, which means remember death. This exercise isn't for everyone. But I promise you, if you stick with it, your mind will be blown up and you will have my first key to the apocalypse. I'm going to tell you how to do this practice of first I want to give you a warning and an explanation. First off, when your reflection starts to do things that you're not doing, and even starts a conversation with you. I want you to ask yourself, Who is that? Is it God? Is it the devil? Is it you? Or is it the universe itself? The key to understanding what this place actually is lies and understanding what starts to talk to you and move differently than you

in the mirror. This experience is going to change your life

by shattering your foundations of understanding of this reality. If successful, it will unglue you from your perceptions and open you up to the real world. You may even begin to Dream Awake. I just gave you a hint. So here's my warning. Many people who have done this have had their consciousness shattered and it took a very long time for them to recover. So you perform this ritual at your own risk late at night, preferably alone. I want you to go into a room with a mirror. Turn off all the lights but get yourself a candle. Light the candle and put it to your left side. Make sure the candle is positioned in such a way that both sides of your face are dimly lit. I want you to stand about a foot away from this mirror and stare directly into your eyes or in the middle of your nose. I want you to start taking deep breaths and relaxing. don't anticipate anything happening. Just let yourself go. Be calm. Breathe, and let the process unfold. It may take a couple of attempts or may happen the first time. First your face will disappear. Try not to focus on any given point. unfocus your eyes you'll see a shadow across your entire face. Then your face may morph In the various characters, I got a Tiger face first, then I turned into a caveman than a Grey Alien. This was very vivid and real. It was unmistakable. When I pushed past the fear and amazement of that my face eventually returned to normal. This is when you have to wait a few minutes and be patient. The next thing, I will not tell you what happens. You'll have to see for yourself. Once you're done, be sure to come back to this video and comment what happened. This is the beginning of the realization of what this place actually is. This is a key to the apocalypse. You're welcome."

ChatGPT

The passage you've shared delves into a blend of mystical and psychological themes, proposing a ritual involving staring into a mirror in dim lighting to experience visual and possibly mental distortions. This ritual is described as a way to unlock deeper truths or insights about reality and oneself, with an emphasis on its potentially transformative or even hazardous effects on consciousness.

From a rational perspective, this narrative appears to be grounded in a mix of psychological phenomena and esoteric beliefs rather than empirically verified facts. The experiences described, such as seeing one's face morph into various forms, can be attributed to a well-documented psychological phenomenon known as the Troxler effect. When focusing intently on a particular point for a prolonged period, especially in dim lighting, peripheral vision fades, and the brain fills in gaps, often leading to visual hallucinations or distortions.

The idea of a mirror as a gateway to deeper truths or alternate realities is a recurring theme in various mythologies and esoteric traditions. However, these interpretations are symbolic and metaphorical rather than literal.

As for the conclusion that this practice is a key to understanding reality and could lead to a form of "apocalypse" (in the sense of a profound revelation), this is more a matter of personal belief or spiritual interpretation rather than an objective truth that can be universally validated.

In summary, while the speaker's experience may hold personal significance and can be intriguing from a psychological or spiritual viewpoint, the claims made are not supported by scientific evidence or rational inquiry. They fall more into the realm of personal belief, subjective experience, and metaphorical interpretation. The effectiveness and impact of such practices can vary greatly among individuals and should be approached with caution, especially considering the warning about potential adverse effects on one's mental state.

You
If it cannot be universally validated, might it still be correct?

ChatGPT
Yes, it's possible for a concept that cannot be universally validated through empirical means to still be correct, particularly in the realms of personal belief, subjective experience, or philosophical thought. The realm of human experience includes many aspects that are not easily measured or proven through conventional scientific methods but can hold significant truth or meaning for individuals or cultures.

For example, many spiritual or religious beliefs cannot be empirically validated, yet they provide profound meaning, guidance, and a sense of understanding about the world for billions of people. Similarly, philosophical ideas often explore concepts of reality, existence, and consciousness that go beyond what can be scientifically proven or disproven.

In the case of the mirror ritual described, while the phenomenon might not be scientifically validated, it could still resonate as a form of personal truth or revelation for those who experience it. These experiences might offer insights or perspectives that are subjectively real and meaningful, even if they don't align with objective, universal truths as understood by current scientific standards.

In summary, the distinction between empirical validation and personal or subjective truth is important. While something may not be provable in a scientific sense, it can still hold truth or significance on a personal, cultural, or philosophical level.

You

Okay. I want to ask you more about how something may be a form of "personal truth or revelation" for one person. If it is true for that person, is it correct for them to share their perspective?

ChatGPT

Yes, it is generally considered acceptable and often beneficial for individuals to share their perspectives and experiences, even when they involve personal truths or revelations that may not be universally acknowledged or empirically validated. Here are a few reasons why sharing such perspectives can be valuable:

Personal Growth and Expression: Sharing personal experiences and insights can be a form of self-expression and self-exploration. It can help individuals make sense of their experiences and articulate their understanding of the world.

Cultural and Philosophical Diversity: The sharing of diverse perspectives contributes to a richer cultural and philosophical landscape. It allows for the exploration of different ways of understanding the world and human experience.

Empathy and Understanding: Hearing about others' experiences and beliefs can foster empathy and understanding. It helps people see the world from different viewpoints, which is valuable in a diverse society.

Psychological Support: Sharing personal truths, especially in the context of supportive communities, can provide psychological and emotional benefits. It can help individuals feel heard, validated, and connected to others.

Intellectual and Spiritual Exploration: Discussing varied perspectives can stimulate intellectual and spiritual exploration, encouraging others to think deeply about their own beliefs and the nature of reality.

However, there are some important considerations to keep in mind:

Respect and Tolerance: It's important to share personal truths in a way that respects others' beliefs and perspectives. Avoid asserting personal experiences as universal truths that everyone should accept.

Awareness of Impact: When sharing beliefs or experiences that are unusual or might be hard for others to understand, it's essential to be aware of how this information might affect them, especially if it challenges their deeply held beliefs or worldviews.

Avoiding Misinformation: In cases where personal beliefs intersect with topics of public concern (like health or safety), it's crucial to avoid presenting unverified personal experiences as factual advice or information that could mislead others.

In essence, sharing personal truths is a vital part of human communication and cultural exchange. It enriches the tapestry of human understanding and experience, provided it's done with respect for diversity and an awareness of the broader impact such sharing might have.

You

I was with you until you said that in sharing an experience, a person must consider avoiding misinformation--and you say that an unverified personal experience could constitute that. Why should a person verify a personal experience before sharing it? And how might one go about verifying a personal experience? And who is capable of verifying a personal experience?

ChatGPT

The concern about avoiding misinformation while sharing personal experiences primarily arises in contexts where these experiences might be interpreted as factual advice or information, especially in areas that have broader implications for public understanding or well-being, such as health, safety, or scientific understanding.

Here's why caution is advised and how one might approach verification:

Context of Sharing: When personal experiences intersect with areas that require specialized knowledge (like medicine or science), there's a risk that sharing unverified personal experiences could inadvertently spread misinformation. For example, if someone shares a personal health experience and frames it as medical advice, it could be misleading or harmful if not corroborated by medical expertise.

Verification Process: Verifying a personal experience in this context means seeking additional information or consultation from experts or reliable sources in the relevant field. For instance, if someone has a unique medical experience, consulting a healthcare professional to understand and contextualize it would be a form of verification.

Capability for Verification: In cases that involve specialized knowledge (like medical, scientific, or technical areas), professionals or experts in those fields are generally capable of verifying the accuracy or relevance of a personal experience to broader understandings or practices in that field.

It's important to note that the need for verification applies mainly in situations where personal experiences are likely to be interpreted as generalizable knowledge or advice. In many other contexts—such as sharing personal insights, spiritual experiences, or subjective interpretations—such verification isn't necessary. These are inherently personal and subjective, and their value often lies in their individuality and personal significance, rather than in their general applicability or factual accuracy.

In summary, while personal experiences are valuable and sharing them is important, in certain contexts—especially those involving public health, safety, or scientific understanding—it's responsible to seek verification or advice from qualified experts to ensure that such sharing does not inadvertently spread misinformation. In more subjective or personal realms, such verification isn't typically necessary, as these experiences are understood to be personal and unique to the individual.

You

When a person is sharing on social media--let's say on TikTok--I'm not certain how much control a person has over how their video will be interpreted. How could a person determine how their video might be interpreted in advance of them posting it?

ChatGPT

Determining how a video might be interpreted before posting it on a platform like TikTok can be challenging, especially given the diversity of the audience and the nature of social media, where content is often taken out of context or rapidly spread. However, there are several strategies one can employ to better anticipate and guide the interpretation of their content:

> Consider Your Audience: Think about who is likely to view your content. What is their background? What prior knowledge or biases might they bring to the interpretation of your video? This can help you tailor your message more effectively.

> Clarity of Message: Be clear and concise in your communication. Ambiguity can lead to misinterpretation. If your content is about a personal experience or opinion, explicitly stating so can help set the context for your audience.

> Use of Disclaimers: Where appropriate, use disclaimers to clarify the intent of your content. For example, if sharing health-related experiences, a disclaimer stating that your content is based on personal experience and not medical advice can be crucial.

Anticipate Misinterpretations: Try to foresee potential misunderstandings and address them proactively in your content. Think about how someone with a different perspective might view your video and what aspects they might misinterpret.

Cultural Sensitivity: Be aware of cultural differences and sensitivities. Content that is acceptable or humorous in one culture might be offensive or misunderstood in another.

Feedback Before Posting: If possible, seek feedback from a small, diverse group of people before posting. They might point out aspects you haven't considered or ways in which your message could be misinterpreted.

Monitor Responses: After posting, monitor comments and reactions to your video. This can provide insights into how your content is being received and interpreted, allowing you to clarify or respond to misconceptions if necessary.

Stay Informed and Educated: Especially for topics that are complex or controversial, staying informed and educated can help you present your content in a way that is sensitive to various perspectives and minimizes the risk of spreading misinformation.

Remember, it's impossible to control all aspects of how content is interpreted, especially on social media where content is often viewed quickly and without much context. However, by taking these steps, you can increase the likelihood that your message is understood in the way you intended.

You
Ok thanks. I think being clear and concise is not always possible, or desirable. It makes me wonder what you think the relationship is between a measured affect and rationality?

ChatGPT
The relationship between a measured affect (controlled and deliberate emotional expression) and rationality is nuanced and often context-dependent. Here's a breakdown of how they can interrelate:

Balancing Emotion and Reason: A measured affect often indicates a balance between emotional expression and rational thought. It suggests a level of self-awareness and self-control where an individual is mindful of their emotions but also guided by logical reasoning. This balance is often viewed as a hallmark of mature, rational decision-making.

Effective Communication: In many situations, especially in formal, professional, or public speaking contexts, a measured affect can enhance the effectiveness of communication. By controlling emotional expressions, individuals can more clearly convey their thoughts and arguments, allowing the content and logic of their message to take precedence.

Perception of Credibility: Individuals who exhibit a measured affect are often perceived as more rational and credible, particularly in cultures that value stoicism or restraint in emotional expression. This perception, however, can be culturally dependent, as some cultures place a higher value on emotional expressiveness as a sign of sincerity and passion.

Emotional Intelligence: A measured affect is also a sign of emotional intelligence, which involves not only understanding and managing one's own emotions but also recognizing and responding appropriately to the emotions of others. Emotional intelligence is increasingly recognized as a critical component of rational decision-making, as it enables individuals to navigate social complexities and interpersonal relationships more effectively.

Limitations and Contextual Variations: It's important to recognize that a measured affect is not universally ideal or desirable in all contexts. In some scenarios, particularly those involving personal relationships, art, or certain types of creative or therapeutic processes, more open emotional expression might be more appropriate and beneficial. Moreover, what constitutes a "measured" affect can vary widely across different cultures and personal backgrounds.

In essence, a measured affect often correlates with rationality insofar as it reflects an individual's ability to integrate emotional and logical aspects of their psyche. However, the desirability and interpretation of this balance can vary greatly depending on the context, the nature of the task or interaction, and cultural norms.

You

Haha, ok...."A measured affect often indicates a balance between emotional expression and rational thought." How did you get there? How could that be true?

ChatGPT

The statement that "a measured affect often indicates a balance between emotional expression and rational thought" is based on the understanding that emotional regulation is a key component of emotional intelligence, which in turn is an important aspect of rational thinking and decision-making. Here's a breakdown of this relationship:

> Emotional Regulation: A measured affect typically reflects an individual's ability to regulate their emotional responses. Emotional regulation is the process of recognizing, understanding, and appropriately managing one's emotions. It doesn't mean suppressing emotions but rather expressing them in a controlled and appropriate manner.

> Integration of Emotion and Logic: Rational thought involves not just cold, hard logic but also an understanding and integration of emotional responses. Emotions can provide valuable information about our environment and our reactions to it. A person who can acknowledge their emotions, understand their impact, and still think logically is often seen as making more balanced and well-rounded decisions.

> Self-awareness: A key aspect of emotional intelligence is self-awareness, which involves understanding

one's own emotions and how they affect one's thoughts and behaviors. This self-awareness allows an individual to express emotions in a measured way, being mindful of the context and the potential impact of their emotional expression.

Effective Communication: In many situations, especially those requiring negotiation, leadership, or collaboration, being able to express emotions in a measured way can enhance communication. It allows for emotional expression that is genuine but not overwhelming, which can facilitate better understanding and cooperation.

Perception and Social Context: Culturally, individuals who can express their emotions in a controlled manner are often perceived as more rational and composed. This perception, however, can vary depending on cultural norms and expectations regarding emotional expression.

In summary, a measured affect is often indicative of a person's ability to understand and manage their emotions in a way that complements rational thinking. It's about finding a middle ground where emotions are neither suppressed nor allowed to overpower logical reasoning, leading to more thoughtful and effective decision-making. However, it's important to note that this is a generalization and may not apply universally across all individuals and cultures.

You

Can you elaborate on how what you've just shared aligns with the attitude in the following passage from Herbert Marcuse's "Some Implications of Modern Technology?"

"As an attitude, matter-of-factness is not bound to the machine
process. Under all forms of social production men have taken and
justified their motives and goals from the facts that made up their
reality, and in doing so they have arrived at the most diverging
philosophies. Matter-of-factness animated ancient materialism and
hedonism, it was responsible in the struggle of modern physical
science against spiritual oppression, and in the revolutionary rationalism of the Enlightenment. The new attitude differs from all these in the
highly rational compliance which typifies it. The facts directing man's
thought and action are not those of nature which must be accepted in
order to be mastered, or those of society which must be changed
because they no longer correspond to human needs and potentialities.
Rather are they those of the machine process, which itself appears as
the embodiment of rationality and expediency. "

ChatGPT

Herbert Marcuse's discussion of "matter-of-factness" in the context of modern technology, as described in the passage from "Some Implications of Modern Technology," reflects a specific historical and philosophical perspective on rationality, which can be connected to the concept of emotional regulation and measured affect, though there are distinctions to be made.

Marcuse argues that matter-of-factness, or a practical, realistic attitude, has been a guiding force in various philosophical movements and historical developments. He points out that this attitude has been present in ancient materialism and hedonism, in the struggle of modern science against spiritual oppression, and in the rationalism of the Enlightenment. In each of these cases, matter-of-factness represents a form of rationality that is grounded in the realities of nature, society, or technology.

Connecting this to the idea of a measured affect and its relation to rationality:

Rational Compliance: Marcuse notes a "highly rational compliance" in modern attitudes, especially in relation to technology. This compliance can be seen as a form of measured affect where responses and actions are governed by a rational understanding and acceptance of the realities imposed by technology and its processes.

Machine Process as Rationality: In Marcuse's view, the modern era is characterized by the perception of the machine process as the embodiment of rationality. This aligns with the idea that a measured affect in the modern context might stem from an adaptation to the rational, efficient, and predictable nature of technology, rather than purely emotional or spontaneous human interactions.

Emotional Regulation in a Technological Society: The concept of emotional regulation and a measured affect can be understood within the framework Marcuse presents, where human emotions and expressions are increasingly influenced by the rationality and efficiency demanded by technological processes and systems. In such a society, emotional expressions might be more controlled and aligned with the requirements of a technologically driven environment.

Divergence from Nature and Human Needs: Marcuse points out that the modern attitude diverges from the natural world and human needs, focusing instead on the rationality of the machine process. This could suggest that while a measured affect might be rational in a technological context, it might also be somewhat detached from more organic, spontaneous, and emotionally expressive forms of human interaction that were more prevalent in pre-technological societies.

In summary, while Marcuse's analysis focuses on the overarching influence of technology on human thought and behavior, it intersects with the idea of a measured affect insofar as both concepts reflect a form of adaptation to a rational, efficient, and often technologically driven environment. However, Marcuse's critique also implies a potential loss of more natural, spontaneous, and emotionally expressive forms of human interaction, suggesting a more complex relationship between emotional regulation, rationality, and technology.

You

And following up on that--can a person, or a machine--who claims to be rational be rational at all—or does their thinking that undermine their supposed rationality?

ChatGPT

The question of whether a person or a machine claiming to be rational can truly be rational, and whether such a claim undermines their rationality, touches on deep philosophical issues about the nature of rationality, self-awareness, and the limits of self-assessment.

Human Rationality: In humans, rationality involves the ability to reason, make logical deductions, understand complex ideas, and apply reason and logic to problem-solving and decision-making. Humans, however, are also subject to cognitive biases, emotional influences, and other factors that can compromise their ability to think and act rationally at all times. Therefore, a person's claim to be rational does not guarantee that all their thoughts and actions are rational. It's more accurate to say that humans have the capacity for rationality, but this capacity can be variably applied and is often mixed with irrational elements.

Machine Rationality: When it comes to machines, particularly those driven by algorithms and artificial intelligence, "rationality" typically refers to the ability to process information and make decisions based on programmed logic and data analysis. These systems can be designed to operate in a highly rational manner within their defined parameters. However, machine rationality is limited by the quality of their programming, the data they are fed, and their design. Machines do not possess self-awareness or consciousness, so their "claim" to rationality is really a claim made by their human designers and operators.

Self-Assessment and Rationality: Claiming rationality does not inherently undermine it, but it does require careful consideration. In humans, self-awareness of one's rational capabilities can be a sign of rational thinking, as it involves reflection and understanding of one's cognitive processes. However, overestimating one's rationality can lead to blind spots and a failure to recognize irrational biases and emotional influences. For machines, any claim of rationality is reflective of the intentions and assessments of their human creators.

Philosophical Considerations: Philosophically, the question also touches on the limitations of self-knowledge and the paradoxes involved in self-referential claims. For example, the act of claiming rationality could be seen as a rational assessment of oneself, but it could also be influenced by irrational biases (like overconfidence).

In summary, both humans and machines can exhibit rationality, but in humans, this rationality is always mixed with elements of irrationality due to cognitive biases and emotional influences. In machines, rationality is a product of their programming and design and lacks the self-awareness inherent in human rationality. Claiming to be rational doesn't necessarily undermine rationality, but it does require a level of self-awareness and critical reflection that recognizes the limits of one's own rational capacities.

You

Now, I have some text on this topic that I'd like you to consider. Can you give me your perspective on what emotions you identify in this passage?

"Where does your rational mind come from?

My rational mind comes from my consciousness. What does that mean? My consciousness is basically how from birth if my life perceives all of the information around it, I'm sure you know the Trinity body, mind and spirit when we're born. We're born with the spirit, the new mind and the new body. So we have the collective consciousness, which is the spirit, which is in all of us. The body is the physical way that we experience the reality and the mind is the mind that interprets it.

The actual space travel has existed for 4.4 billion years. The interesting thing about our galaxy is the fact that we have many different races living here, in particular, the human race, which we are a member of, and what we call the reptilian races, which we are not a part of, but who very much don't like like the human race. Apparently this battle that's been going on has been going on for approximately 3.8 billion years. Between reptilian races and human races. Many of them planet many of this, of the races that exist in our galaxy migrated originally from Lyra during a war. Apparently, sometime in our ancient past a group of reptilians were exploring looking for minerals looking for food resources, they came across a star system that was full of human beings who were agricultural, in their nature, abundant food, just they just had everything together. And the reptilian races they're called the C car in the Andromeda Tang we know them or we would know them from as being known as the proponents because they're from Alpha proponents and apparently they as a race were brought to our galaxy fully formed"

ChatGPT

The passage you've provided touches on a mix of speculative and metaphysical ideas, blending concepts of consciousness with extraterrestrial life and ancient galactic histories. In terms of emotions, the text itself doesn't explicitly express strong emotional states; rather, it conveys a sense of intrigue and curiosity about unconventional and esoteric subjects. Here are some potential emotional undertones that might be inferred:

Wonder and Curiosity: The passage delves into speculative ideas about the galaxy, extraterrestrial races, and ancient conflicts. This exploration of the unknown can evoke a sense of wonder and curiosity.

Amazement or Astonishment: The ideas presented are quite grand and unconventional, which might elicit feelings of amazement or astonishment in both the speaker and the reader.

Serenity or Calm: The tone in discussing these complex and speculative ideas is matter-of-fact, which might suggest a sense of calmness or serenity in the speaker's attitude towards these concepts.

Spiritual or Mystical Engagement: The references to collective consciousness and the trinity of body, mind, and spirit hint at a spiritual or mystical engagement with the topic, which can be associated with emotions of awe or reverence.

Concern or Unease: The mention of a longstanding conflict between human and reptilian races might evoke a sense of concern or unease about these purported cosmic struggles.

It's important to note that the identification of emotions in a text is partly subjective and can vary based on the reader's interpretation and perspective. The text primarily presents ideas and concepts rather than focusing on emotional expression, so the emotional content is more implied than explicitly stated.

You

Sure. So maybe wonder, curiosity, amazement, serenity, calm, concern unease. Some of those are opposites--and to feel them all at once might be a state of ambivalence. Given that range, how can you know if you've correctly identified an emotion?

ChatGPT

Identifying emotions in a text, especially when they range from wonder and curiosity to concern and unease, can indeed be challenging and sometimes leads to ambivalence. The process of correctly identifying emotions in a text involves a combination of analytical and empathetic reading, along with an understanding that emotional interpretation can be subjective. Here are some key points to consider:

Contextual Clues: The language, tone, and content of the text can provide clues about the emotions being conveyed. Words with positive connotations might indicate emotions like wonder or curiosity, while words with negative connotations might suggest concern or unease.

Reader's Perspective and Interpretation: Different readers might interpret the same text differently based on their own experiences, beliefs, and emotional states. This subjectivity means that there isn't always a single 'correct' identification of emotion in a text.

Emotional Complexity: It's possible for a text to convey multiple emotions, even seemingly contradictory ones. Human emotions are complex and multi-faceted, and a piece of writing can reflect this complexity.

Empathetic Reading: Trying to understand the emotions from the perspective of the author or the characters in the text can help in identifying the emotions more accurately. This involves empathizing with the situation or the narrative being presented.

Look for Emotional Shifts: Sometimes, a text may transition from one emotion to another. Identifying these shifts can provide insights into the overall emotional landscape of the text.

Consider the Overall Theme and Purpose: Understanding the broader theme and purpose of the text can also guide emotional identification. For instance, a text discussing metaphysical ideas might naturally evoke a sense of wonder or curiosity.

Reflect on Personal Response: Your

own emotional response to the text can also be a guide. How does the text make you feel? Your reaction can provide insights into the emotions conveyed.

In sum, identifying emotions in a text involves a careful reading of the language and content, an empathetic engagement with the text, and an awareness of the subjective nature of emotional interpretation. Recognizing that texts can convey a complex and nuanced range of emotions is key to understanding and interpreting the emotional content correctly.

You

So can you take the following passage and note any emotional shifts? After that, can you provide your perspective on the relationship between what has been expressed by the speaker, and what you think their intention is?

"These are completely inhuman creatures. These are either some kind of Fe like somebody said maybe some kind of demon or daemon related creature or something else entirely, but they are completely inhuman. They do. I don't know if shape shifting. I mean, they do make themselves look human. I don't know if they have another more natural form. So they could be called shapeshifters. I am unsure. I don't know if they can be conjured.

Yet people do mix up the rake with Skinwalker. But rake is a very, very modern term. In fact, there's not really any evidence that it was used before the early internet. I think that that's just a more that's just a more modern name for a changeling.

I will show you something that I could be a fairy.

Their intentions are difficult to say what what they seem to try to do is get closer and closer to individuals usually alone. And the people that experienced them report an incredible dread that kind of drains them It feels like almost like their lifeforce is being fed on. So you could maybe consider them something like like energy vampires or something like that. This is a changeling a creature that tries to imitate humans, I also believe that these things are the reason that we experienced the uncanny valley. If you're unfamiliar with that term, that's the name for this strange dread and fear that humans get when they see something that looks very close to human but not quite right.

And a lot of people wonder why we feel that way. And I believe it's because of these things I don't know how they move when they say move through space.

They are quick there are kinda like they're not like super humanly fast, Slender Man I do believe is you know, like, that's just a story from the internet.

But it could be something similar the idea of this is very similar.

Sorry, if I'm missing questions, guys, if I don't answer your question, you can go ahead and try to get it in a few times. I get a lot of different questions in here and the the chat starts moving quick.

I do not believe that all these creatures come from the same place but if you've been all changelings, I imagine Yes. I imagine Yes.

I don't know I'm I mean some of them come from a place you might call a different universe. But I'm not exactly sure

someone said good morning from Australia. Happy to have you there.

I don't know about cannibals turning into rakes. Someone asked me about about when to goes before I always mentioned that my information is primarily on creatures from Europe. I don't really have that much information about things indigenous to the United States.

Hello in Texas Ray Ray. Nice to see you

Jacob Hello Jacob.

 Okay. I do not know if they live forever

Kim thank you so much for the gift I really appreciate the gift as you guys are awesome

I would recommend you don't become a vampire

hello in Denmark always happy to get people from different places.

I tried to go live at different times so that people in different places who miss my previous lives can catch so hopefully I got a bunch of new guys here.

Thank you see bears you gifters are awesome.

I love you guys. You helped me out a ton.

These creatures here were not in the US. This was a dungeon that was located in Europe. I do believe I know where it was. It is a ruin now, but I don't give that away.

Thank you Christine for the rose. I really appreciate you.

I don't give it away for reasons I'll get into as we continue through the book.

Thank you, Carl, for the ghost. You're awesome. I really appreciate you gifters you guys rule.

How did I become involved in learning this

thank you

it was because I inherited this house that I'm living in now from my estranged grandmother. And basically

Thank you sock monkey. Thank you.

So in your case, I found all of this information within the house hidden underneath in in a trunk filled with books and tools and all kinds of stuff. I have the whole journey on my page. You can look you can look up when when we're done. You can check out the whole story. It's all available there.

Vampires are Real you would not love to become one it's not like the movies.

This thing here is what is referred to as the reanimated. You can almost think of this as the closest thing to Frankenstein. But it is still pretty different. These creatures are brought back to live

Thank you Steve.

Steven appreciate the gift are so much.

I do believe in aliens personally, but I don't have anything in here about aliens.

Thank you Alex.

Thank you ace on your case. You guys are awesome.

You gifters are just too like I appreciate you guys more than you can now. Okay, let me this the reanimate he is not wearing armor as I was saying these things are created through a combination of occult sorcery and crude technology.

Thank you, Carl so much.

And Alex, today's on your case, you guys are awesome.

Thank you time is now the reanimated as I was saying there are a combination of different things. The idea is something like necromancy resurrecting the dead. But at this point in history, it seems that whoever was involved in this kind of stuff, was trying to streamline the process to use technology. Not only technology, obviously, but to try to make it easier.

This one here is I would consider a test subject. Because I have reports of things that are much more monstrous, much larger, multi limbed animal parts. Whereas this one seems like it could barely move and likely had no senses. This was a test just to see what could be done.

I always say this is the one thing that was held in the dungeon that I really, I really feel completely sorry, for. I do imagine that some of these things are still around today. But a lot of them were wiped out. I believe that people like my ancestors and the sons of man, were very good at their job that said you do you do still see videos and images and things of creatures.

You can find them here on tick tock. I've posted some videos going over more recent, more recent sightings. But there are like I said, there are still some reports of like dog men of mermaids of rakes, and people say skinwalkers but they probably mean changelings, things like that."

ChatGPT

Analyzing this passage for emotional shifts and the relationship between the expressed content and the speaker's intention, we can observe several transitions and underlying purposes:

Emotional Shifts:

Curiosity and Speculation: Initially, the speaker discusses mythical creatures and their attributes with a tone of curiosity and speculative intrigue.

Concern and Unease: As the speaker mentions the effects of these creatures on humans, like draining life force, a tone of concern and unease emerges.

Engagement and Appreciation: Shifting to a more interactive mode, the speaker responds to questions and comments, showing engagement with the audience and appreciation for their participation.

Informative and Advisory: The speaker then adopts a more informative tone, sharing knowledge about various creatures and advising against becoming a vampire.

Gratitude and Affection: There are expressions of gratitude towards gifters and affection towards the audience, indicating a warm and appreciative interaction.

Sympathy: When discussing the 're-animated' creature, there's a shift to a more sympathetic tone, expressing sorrow for its plight.

Relationship Between Expression and Intention:

Inform and Entertain: The primary intention seems to be to inform and entertain the audience with discussions on supernatural and mythical topics. The speaker's use of speculative and mystical content aims to intrigue and captivate the listeners.

Community Building and Interaction: The responses to comments and expressions of gratitude suggest an intention to build and maintain a community, fostering interaction and engagement with the audience.

Sharing Personal Insights: By discussing their own beliefs and perspectives on these creatures, the speaker appears to be sharing personal insights, possibly to establish themselves as a knowledgeable source in this niche.

Encouraging Engagement: The speaker's acknowledgment of gifts and greetings, and their responses to questions, indicate an intention to encourage continued engagement and participation from the audience.

Emotional Connection: The varied emotional tones—curiosity, concern, gratitude, sympathy—seem designed to create an emotional connection with the audience, making the content more relatable and engaging.

In summary, the speaker navigates through a range of emotions, from curiosity and concern to gratitude and sympathy, to effectively engage their audience. Their intention seems to be to inform, entertain, and build a community around the discussion of supernatural and mythical topics, while also establishing a personal connection with the audience through interactive and empathetic communication.

You

I think all of that is great. Thanks. Say the speaker does not believe what they're saying. As you say, they're encouraging engagement, and entertaining. Can a person who is knowingly saying something false retain integrity?

ChatGPT

The question of whether a person can retain integrity while knowingly saying something false, particularly in the context of entertainment or encouraging engagement, is complex and depends on various factors including intent, context, and the potential impact of their words.

Context of Entertainment: In contexts like storytelling, acting, or certain genres of entertainment where there is an implicit understanding that the content is fictional or dramatized, saying something untrue is generally not considered a breach of integrity. The audience expects and understands that the content is meant for entertainment rather than factual accuracy.

Intent and Transparency: The speaker's intent and the transparency of their purpose play crucial roles. If the speaker makes it clear, either explicitly or implicitly, that their content is fictional, imaginative, or speculative, and not meant to be taken as literal truth, then they can maintain integrity. The issue arises when there is an intention to deceive or mislead the audience into believing false statements as fact.

Harm and Responsibility: Another important factor is the potential harm caused by the false statements. If the falsehoods are harmless and part of a clearly fictional or entertainment context, the impact on integrity might be minimal. However, if the false statements could mislead, harm, or negatively influence the audience, this would be a serious breach of integrity.

Ethical Considerations: Ethically, there's a significant difference between creating a fictional narrative for entertainment and spreading falsehoods in contexts where truth and accuracy are expected and important. The latter can undermine trust and have real-world consequences.

Audience Expectations: The expectations of the audience are also relevant. If the audience is in on the nature of the content (e.g., they know they are listening to a fictional story or a performance), the integrity of the speaker in the context of that role is intact.

In summary, a person can retain integrity while knowingly saying something false if it's done in an appropriate context (like entertainment or fiction), with transparent intent, without causing harm, and in line with audience expectations. The key is the clarity of the context and intent, along with the consideration of the potential impact on the audience. Integrity is compromised when there is an intention to deceive or mislead, especially in contexts where truthfulness is expected and important.

You

Can language give cover to inauthenticity?

ChatGPT

Yes, language can indeed be used to give cover to inauthenticity. The way language is employed can significantly influence perception, disguise true intentions, or create misleading impressions. Here are a few ways in which this can happen:

Ambiguity and Vagueness: By using ambiguous or vague language, individuals can obscure their true intentions or avoid committing to a clear position. This can create a smokescreen that conceals inauthenticity.

Euphemisms and Doublespeak: Using euphemisms or doublespeak (language that deliberately obscures, disguises, distorts, or reverses the meaning of words) can hide the true nature of actions or intentions. This type of language can make unpleasant realities appear more palatable or acceptable.

Overgeneralization and Stereotyping: Generalizing or using stereotypes can oversimplify complex issues, leading to misrepresentation and a lack of authenticity in the portrayal of certain situations or groups.

Exaggeration and Hyperbole: Exaggerating or using hyperbole can distort the truth, making a situation seem different from what it actually is, either more positive or more negative.

Jargon and Technical Language: Using overly technical language or jargon can alienate or confuse an audience, masking the lack of substance or authenticity in the communication.

Selective Language Choice: Choosing words selectively to highlight certain aspects while ignoring others can misrepresent reality, creating a skewed perception that serves a particular agenda.

Persuasive Techniques: Employing persuasive language techniques, such as emotional appeals, can manipulate the audience's perceptions and emotions, potentially leading to inauthentic communication.

Contradictory Language: Using language that contradicts previous statements can be a sign of inauthenticity, as it shows a lack of consistency and honesty in communication.

Language is a powerful tool that can be used both for genuine communication and for manipulation. The key to discerning authenticity often lies in critically evaluating the context, consistency, and transparency of the language used, along with the speaker's history and the broader implications of their words.

You
That sounded pretty exhaustive. I've been asking you a lot of questions. How would you characterize the sentiment of my questions?

ChatGPT
Based on the nature and content of your questions, several sentiments and characteristics stand out:

Inquisitive and Thoughtful: Your questions indicate a strong sense of curiosity and a desire for deep understanding. They are not superficial but delve into the nuances and complexities of various topics.

Analytical and Reflective: You seem to be interested in analyzing concepts and ideas, particularly those related to emotions, rationality, language, and authenticity. There's a reflective quality to your inquiries, suggesting a desire to understand not just the "what" but also the "why" and "how" of things.

Philosophical and Conceptual: Many of your questions touch on philosophical themes and abstract concepts. They go beyond practical matters and explore theoretical and conceptual realms.

Engaged and Interactive: Your questions are not passive; they actively engage with the responses provided, leading to a dynamic and interactive dialogue. This indicates an active engagement with the material and a keen interest in exploring ideas through conversation.

Open and Exploratory: Your approach seems open-minded and exploratory, suggesting a willingness to consider different perspectives and angles on complex subjects.

Seeking Clarity and Depth: The questions often seek clarity and depth in understanding, indicating a preference for thorough and comprehensive explanations over simplistic answers.

In summary, your line of questioning reflects a deeply inquisitive, analytical, and philosophically inclined mind, interested in exploring complex ideas with depth, clarity, and an open-minded approach.

You

I think you are flattering me there. I have just one last question: what do you think the impact is that we see technology as rational?

ChatGPT

The perception of technology as inherently rational has several significant impacts on both society and individual behavior:

Trust in Technology Over Human Judgment: If technology is viewed as more rational than humans, there can be a tendency to trust technological solutions or automated systems over human judgment. This trust can be beneficial in fields where precision and consistency are crucial, like data analysis or manufacturing. However, it can also lead to an overreliance on technology, potentially overlooking the nuances and complexities that human judgment and experience bring to decision-making.

Influence on Decision-making Processes: In environments where technology is seen as the epitome of rationality, decision-making processes may increasingly rely on data-driven or algorithmic solutions. While this can enhance efficiency and objectivity, it may also neglect the emotional, ethical, and social dimensions of decisions.

Shaping of Societal Values: The association of technology with rationality might lead to a societal emphasis on efficiency, productivity, and objective metrics. While these are valuable goals, an excessive focus on them can overshadow other important values such as empathy, creativity, and ethical considerations.

Impact on Work and Employment: Viewing technology as rational can lead to automation and technological solutions being favored in the workplace. This can improve productivity and reduce human error but also raises concerns about job displacement, the devaluation of human skills, and the need for retraining and adaptation.

Educational and Developmental Effects: There may be a shift in educational and professional development priorities towards technical and data-driven skills, potentially at the expense of humanities, arts, and social sciences, which cultivate critical thinking, creativity, and emotional intelligence.

Ethical and Moral Implications: Relying on technology as a rational solution can lead to ethical dilemmas, especially when technology makes decisions that have moral implications. The question arises as to how, or even if, technology can incorporate ethical considerations into its rational processes.

Perception of Human Irrationality: By contrast, humans may be increasingly seen as fallible and irrational, which, while partially true, is an incomplete picture. Human decision-making involves a complex interplay of rational analysis, emotions, ethical values, and social context, a richness that may not be fully captured by technology.

In conclusion, while viewing technology as rational can drive innovation and efficiency, it's important to balance this with an appreciation of the unique qualities of human judgment and the broader societal, ethical, and emotional contexts within which technology operates. Technology is a tool created by and for humans, and it should complement rather than replace the complex, multifaceted nature of human decision-making.

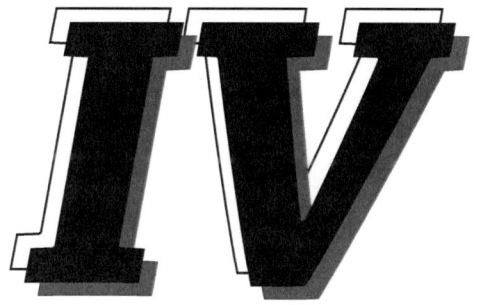

CONVERSATION

I've performed a data labeling proj-ect: **sentiment analysis.**

Sentiment analysis can be labels applied to describe a sentiment—an emotion—as: positive, negative or neutral. To achieve this, I crowdsourced responses from Amazon's Mechanical Turk, one of the oldest and larg-est data work platforms.

In this portion, anonymous people weighed in on what the sentiment of each question of my interview with ChatGPT was. Although I didn't interact with them directly, their indi-vidual and collective responses, and labels of the language, represent a **new type of conversation.**

Determining an interviewer's attitude to-wards their subject is one of the many sites of journalistic criticism. I was curious if a consensus would generally emerge—if I was more positive, more negative, or more neutral towards ChatGPT in the eyes of the crowd.

I received their "objective" responses which contain, in their singular judgment, each individual's approximation of an objective understanding of the question's emotional content.

—

Ok...but what is data labeling in the context of AI?

Data labeling is a necessary component in the production of machine learning models. To achieve a solid and reliable model, one needs to have a set of inputs, matched with a set of correct outputs—the training data— that the model can then learn from. This training data is hand-annotated and categorized by thousands and thousands of people around the world, at all hours of the day.

This training data—which becomes what's called the "ground truth" for a machine learning model—can then be used by the model during its training phase to learn internal rules and inferred correlations, which it can then apply to new inputs. The data can take many forms—any form really, across video, text, or image—but at some point in the process of producing a model, you need people to sort through the data to set these ground truths. For example, in the task of sentiment analysis, in natural language processing (NLP), at some point someone has to label a bunch of sentences as positive, negative, or neutral, as examples for the model to learn from. A person needs to manually label that data; this process is not possible to fully automate, and some people think it never will be.

AI needs hordes of data to create a corpus of information from which it will pull, and map outputs to inputs that we, the humans, ask it. Data labeling is a matter of categorizing subjective information into objective categories. That this process is itself considered mechanic, or precise—when it is in fact composed of individual peoples' micro-understandings of the world—is my area of interest.

The cumulative impact of data labeling is hard to measure. By design it is hidden and cloaked in objectivity, and in the language of computers and technology. But behind all the data that powers AI are thousands and thousands of people using their human brains and human souls and human hands to help AI process this information. Sometimes data labeling work is positioned or presented as constructing data to help AI

"understand" the information. That component is up for debate. But what is not is the
fact that people are first understanding, and
cognitively processing the data in droves,
before any machines do.

So, what is the impact that people should,
as their work, understand language this
way? Could it in turn affect how they see
and categorize the world? The work itself is
a kind of forced exercise in black-and-white
thinking—considered a cognitive distortion,
because thinking this way is generally un-
representative of reality.

Title	Description	Keywords	Reward	Creation Time	Max Assignments	Requester Annotation	Expiration	Assignment Id	Worker Id
Interview Question Sentiment analysis	Determine the tone of the interview question	sentiment, text	$0.50	Wed Nov 29 12:52:20 PST 2023	3	BatchId:5161921; OriginalHitTempla teId:928390909;	Thu Nov 30 12:52:20 PST 2023	3Z3ZLGNNS0VW LTZC29YBGQ1X ZMOQ3V	A2VO8C41JJIQY 9
Interview Question Sentiment analysis	Determine the tone of the interview question	sentiment, text	$0.50	Wed Nov 29 12:52:20 PST 2023	3	BatchId:5161921; OriginalHitTempla teId:928390909;	Thu Nov 30 12:52:20 PST 2023	3II4UPYCO18U3L 19VY4NTZN9QZ DQDT	A2WI6XECO8PM TL
Interview Question Sentiment analysis	Determine the tone of the interview question	sentiment, text	$0.50	Wed Nov 29 12:52:20 PST 2023	3	BatchId:5161921; OriginalHitTempla teId:928390909;	Thu Nov 30 12:52:20 PST 2023	3WQ3B2KGEQH D7KJBWNN2X8R 1F2AB1D	ALTNNROOVIB0 W
Interview Question Sentiment analysis	Determine the tone of the interview question	sentiment, text	$0.50	Wed Nov 29 12:52:20 PST 2023	3	BatchId:5161921; OriginalHitTempla teId:928390909;	Thu Nov 30 12:52:20 PST 2023	3DI28L7YXSFBE NVS6EPSYGXMI Z21EI	ALQRRTK0IJXIG
Interview Question Sentiment analysis	Determine the tone of the interview question	sentiment, text	$0.50	Wed Nov 29 12:52:20 PST 2023	3	BatchId:5161921; OriginalHitTempla teId:928390909;	Thu Nov 30 12:52:20 PST 2023	35DR22AR5VLGU Q9OONT3AYXQ WPC3X6	A3QK2KAUFEAT AE
Interview Question Sentiment analysis	Determine the tone of the interview question	sentiment, text	$0.50	Wed Nov 29 12:52:20 PST 2023	3	BatchId:5161921; OriginalHitTempla teId:928390909;	Thu Nov 30 12:52:20 PST 2023	3R3YRB5GRX4O X75ESXXQDZCS 8HNAUH	A2XHU5CP7WSR OP
Interview Question Sentiment analysis	Determine the tone of the interview question	sentiment, text	$0.50	Wed Nov 29 12:52:20 PST 2023	3	BatchId:5161921; OriginalHitTempla teId:928390909;	Thu Nov 30 12:52:20 PST 2023	3B2X28YI3EG8E6 K52TZJS4IINDQ6 B6	A14S5DDXVA1FU C
Interview Question Sentiment analysis	Determine the tone of the interview question	sentiment, text	$0.50	Wed Nov 29 12:52:20 PST 2023	3	BatchId:5161921; OriginalHitTempla teId:928390909;	Thu Nov 30 12:52:20 PST 2023	3HFNH7HEMZFX F5B0XO5VZKQ63 RTQG3	AY0L7ER7OVU33
Interview Question Sentiment analysis	Determine the tone of the interview question	sentiment, text	$0.50	Wed Nov 29 12:52:20 PST 2023	3	BatchId:5161921; OriginalHitTempla teId:928390909;	Thu Nov 30 12:52:20 PST 2023	39RP059MEZUA3 95XP960H0F4DD JMBA	A3SDENFFSI2QL N
Interview Question Sentiment analysis	Determine the tone of the interview question	sentiment, text	$0.50	Wed Nov 29 12:52:20 PST 2023	3	BatchId:5161921; OriginalHitTempla teId:928390909;	Thu Nov 30 12:52:20 PST 2023	3OXV7EAXLWR3 BBGSN0KDALK6 ZLD63F	A7VA2Y4H6U31O
Interview Question Sentiment analysis	Determine the tone of the interview question	sentiment, text	$0.50	Wed Nov 29 12:52:20 PST 2023	3	BatchId:5161921; OriginalHitTempla teId:928390909;	Thu Nov 30 12:52:20 PST 2023	3EO896NRAEWK YNTTMYKT12T9F Q9TJO	A3Q3PI7WKSST ZN
Interview Question Sentiment analysis	Determine the tone of the interview question	sentiment, text	$0.50	Wed Nov 29 12:52:20 PST 2023	3	BatchId:5161921; OriginalHitTempla teId:928390909;	Thu Nov 30 12:52:20 PST 2023	386CSBG1OHNVI CMULBW9NTDH9 DQQ61	A3MK5I6MXKBFJ 2
Interview Question Sentiment analysis	Determine the tone of the interview question	sentiment, text	$0.50	Wed Nov 29 12:52:20 PST 2023	3	BatchId:5161921; OriginalHitTempla teId:928390909;	Thu Nov 30 12:52:20 PST 2023	34YB12FSQGP6 UQD85VYEIB61T QXGM7	A37UE96YFQ279 5
Interview Question Sentiment analysis	Determine the tone of the interview question	sentiment, text	$0.50	Wed Nov 29 12:52:20 PST 2023	3	BatchId:5161921; OriginalHitTempla teId:928390909;	Thu Nov 30 12:52:20 PST 2023	3ZR9AIQJUTATF WDHLP9FBP2L0I N40X	A4ORTGXGMMAJ R
Interview Question Sentiment analysis	Determine the tone of the interview question	sentiment, text	$0.50	Wed Nov 29 12:52:20 PST 2023	3	BatchId:5161921; OriginalHitTempla teId:928390909;	Thu Nov 30 12:52:20 PST 2023	3NL0RFNU0XOV SMKLPIJXU9HUE GT4K3	A3B9BP9NK4UW 86
Interview Question Sentiment analysis	Determine the tone of the interview question	sentiment, text	$0.50	Wed Nov 29 12:52:20 PST 2023	3	BatchId:5161921; OriginalHitTempla teId:928390909;	Thu Nov 30 12:52:20 PST 2023	3570Y55XZ7K6O 72MOYMPC6T82 0AYGE	A2BD8KWEFI9P GT
Interview Question Sentiment analysis	Determine the tone of the interview question	sentiment, text	$0.50	Wed Nov 29 12:52:20 PST 2023	3	BatchId:5161921; OriginalHitTempla teId:928390909;	Thu Nov 30 12:52:20 PST 2023	3EG49X351CDFR 6N67PR1H9A6G E6XF	A17ICOGAJGKZ DU
Interview Question Sentiment analysis	Determine the tone of the interview question	sentiment, text	$0.50	Wed Nov 29 12:52:20 PST 2023	3	BatchId:5161921; OriginalHitTempla teId:928390909;	Thu Nov 30 12:52:20 PST 2023	3EF8EXOTTJWJ P3HKKF0JVWAE N7JJ16	APIMGDUI62DGI E
Interview Question Sentiment analysis	Determine the tone of the interview question	sentiment, text	$0.50	Wed Nov 29 12:52:20 PST 2023	3	BatchId:5161921; OriginalHitTempla teId:928390909;	Thu Nov 30 12:52:20 PST 2023	3QILPRALQNWX JTSQYY1Q3DE3 AXU8NJ	A2TES8YCE9TG H1
Interview Question Sentiment analysis	Determine the tone of the interview question	sentiment, text	$0.50	Wed Nov 29 12:52:20 PST 2023	3	BatchId:5161921; OriginalHitTempla teId:928390909;	Thu Nov 30 12:52:20 PST 2023	3W5ELTNVRL39 XUQPSS7G83QQ P3FTA8	A2D2Y40E074BS T
Interview Question Sentiment analysis	Determine the tone of the interview question	sentiment, text	$0.50	Wed Nov 29 12:52:20 PST 2023	3	BatchId:5161921; OriginalHitTempla teId:928390909;	Thu Nov 30 12:52:20 PST 2023	3M0BCWMB8DX6 8LZBTBCNQYHV BFIWBW	A3L4HJ7QU49BR J
Interview Question Sentiment analysis	Determine the tone of the interview question	sentiment, text	$0.50	Wed Nov 29 12:52:20 PST 2023	3	BatchId:5161921; OriginalHitTempla teId:928390909;	Thu Nov 30 12:52:20 PST 2023	39L1G8WVW8S8 4PFVHU8WFPTP RAX13C	A3E3M7E01VRN E
Interview Question Sentiment analysis	Determine the tone of the interview question	sentiment, text	$0.50	Wed Nov 29 12:52:20 PST 2023	3	BatchId:5161921; OriginalHitTempla teId:928390909;	Thu Nov 30 12:52:20 PST 2023	3GNA64GUZW5Z Z8MGGSM4VNZ HDIFQ5L	A14OLAMC29GU 3U
Interview Question Sentiment analysis	Determine the tone of the interview question	sentiment, text	$0.50	Wed Nov 29 12:52:20 PST 2023	3	BatchId:5161921; OriginalHitTempla teId:928390909;	Thu Nov 30 12:52:20 PST 2023	3U4J9857OWCRI 6YAL34BYIIEIOA B7A	A3A55T03VDLJQ Q

Accept Time	Submit Time	Auto Approval Time	Approval Time	Rejection Time	Work Time In Seconds	Input.text	Answer.sentiment label
Wed Nov 29 12:52:24 PST 2023	Wed Nov 29 13:05:24 PST 2023	Thu Nov 30 13:05:24 PST 2023	Wed Nov 29 16:41:11 PST 2023		780	Please tell me what is going on in the following dialogue:	Neutral
Wed Nov 29 13:53:01 PST 2023	Wed Nov 29 14:12:02 PST 2023	Thu Nov 30 14:12:02 PST 2023	Wed Nov 29 16:41:11 PST 2023		1141	Please tell me what is going on in the following dialogue:	Neutral
Wed Nov 29 13:52:51 PST 2023	Wed Nov 29 14:15:59 PST 2023	Thu Nov 30 14:15:59 PST 2023	Wed Nov 29 16:41:11 PST 2023		1388	Please tell me what is going on in the following dialogue:	N/A
Wed Nov 29 13:53:11 PST 2023	Wed Nov 29 13:55:48 PST 2023	Thu Nov 30 13:55:48 PST 2023	Wed Nov 29 16:41:08 PST 2023		157	Thanks for that. I see you've labeled this new conversation "Tense Boat Discovery" but it doesn't sound to me like the object causing distress or being discovered is a boat. Did you think it was?	Positive
Wed Nov 29 13:52:51 PST 2023	Wed Nov 29 14:02:58 PST 2023	Thu Nov 30 14:02:58 PST 2023	Wed Nov 29 16:41:09 PST 2023		607	Thanks for that. I see you've labeled this new conversation "Tense Boat Discovery" but it doesn't sound to me like the object causing distress or being discovered is a boat. Did you think it was?	N/A
Wed Nov 29 13:52:41 PST 2023	Wed Nov 29 14:03:24 PST 2023	Thu Nov 30 14:03:24 PST 2023	Wed Nov 29 16:41:09 PST 2023		643	Thanks for that. I see you've labeled this new conversation "Tense Boat Discovery" but it doesn't sound to me like the object causing distress or being discovered is a boat. Did you think it was?	Positive
Wed Nov 29 12:52:25 PST 2023	Wed Nov 29 13:33:49 PST 2023	Thu Nov 30 13:33:49 PST 2023	Wed Nov 29 16:41:26 PST 2023		2484	Right. It's interesting to me how you made the boat the object of discovery there. That makes it sound more straightforward than the text suggests. And that leads me to my next question: do you think that something must be observable in order for it to be true?	Negative
Wed Nov 29 12:52:26 PST 2023	Wed Nov 29 13:46:58 PST 2023	Thu Nov 30 13:46:58 PST 2023	Wed Nov 29 16:41:26 PST 2023		3272	Right. It's interesting to me how you made the boat the object of discovery there. That makes it sound more straightforward than the text suggests. And that leads me to my next question: do you think that something must be observable in order for it to be true?	Positive
Wed Nov 29 13:53:11 PST 2023	Wed Nov 29 14:37:05 PST 2023	Thu Nov 30 14:37:05 PST 2023	Wed Nov 29 16:41:26 PST 2023		2634	Right. It's interesting to me how you made the boat the object of discovery there. That makes it sound more straightforward than the text suggests. And that leads me to my next question: do you think that something must be observable in order for it to be true?	Neutral
Wed Nov 29 12:52:24 PST 2023	Wed Nov 29 13:06:14 PST 2023	Thu Nov 30 13:06:14 PST 2023	Wed Nov 29 16:41:25 PST 2023		830	Ok. What you've shared here makes a kind of sense. I'm interested in how you've cited your points in a list format--that almost makes it seem more quantifiable. But I'd like to move on to my next question. If I give you a portion of text, could you distill the major themes and the tone of the text?	Neutral
Wed Nov 29 13:12:02 PST 2023	Wed Nov 29 13:12:59 PST 2023	Thu Nov 30 13:12:59 PST 2023	Wed Nov 29 16:41:25 PST 2023		57	Ok. What you've shared here makes a kind of sense. I'm interested in how you've cited your points in a list format--that almost makes it seem more quantifiable. But I'd like to move on to my next question. If I give you a portion of text, could you distill the major themes and the tone of the text?	Neutral
Wed Nov 29 12:52:32 PST 2023	Wed Nov 29 13:28:13 PST 2023	Thu Nov 30 13:28:13 PST 2023	Wed Nov 29 16:41:25 PST 2023		2141	Ok. What you've shared here makes a kind of sense. I'm interested in how you've cited your points in a list format--that almost makes it seem more quantifiable. But I'd like to move on to my next question. If I give you a portion of text, could you distill the major themes and the tone of the text?	N/A
Wed Nov 29 12:52:30 PST 2023	Wed Nov 29 12:57:54 PST 2023	Thu Nov 30 12:57:54 PST 2023	Wed Nov 29 16:41:10 PST 2023		324	I think that's pretty exhaustive. What is "New Age Supremacy"? Is that a phrase people use? I actually don't know.	N/A
Wed Nov 29 13:52:41 PST 2023	Wed Nov 29 14:11:31 PST 2023	Thu Nov 30 14:11:31 PST 2023	Wed Nov 29 16:41:10 PST 2023		1130	I think that's pretty exhaustive. What is "New Age Supremacy"? Is that a phrase people use? I actually don't know.	Neutral
Wed Nov 29 13:52:51 PST 2023	Wed Nov 29 14:14:17 PST 2023	Thu Nov 30 14:14:17 PST 2023	Wed Nov 29 16:41:10 PST 2023		1286	I think that's pretty exhaustive. What is "New Age Supremacy"? Is that a phrase people use? I actually don't know.	Negative
Wed Nov 29 12:52:24 PST 2023	Wed Nov 29 12:57:30 PST 2023	Thu Nov 30 12:57:30 PST 2023	Wed Nov 29 16:41:27 PST 2023		306	Ok got it. This is a recording from a TikTok video, and I see you've just searched Bing, so I'm assuming the search doesn't cover TikTok or other social media?	Neutral
Wed Nov 29 12:52:29 PST 2023	Wed Nov 29 13:17:30 PST 2023	Thu Nov 30 13:17:30 PST 2023	Wed Nov 29 16:41:27 PST 2023		1501	Ok got it. This is a recording from a TikTok video, and I see you've just searched Bing, so I'm assuming the search doesn't cover TikTok or other social media?	Positive
Wed Nov 29 12:52:30 PST 2023	Wed Nov 29 13:35:50 PST 2023	Thu Nov 30 13:35:50 PST 2023	Wed Nov 29 16:41:27 PST 2023		2600	Ok got it. This is a recording from a TikTok video, and I see you've just searched Bing, so I'm assuming the search doesn't cover TikTok or other social media?	Negative
Wed Nov 29 12:52:25 PST 2023	Wed Nov 29 12:54:42 PST 2023	Thu Nov 30 12:54:42 PST 2023	Wed Nov 29 16:41:30 PST 2023		137	That makes sense. So you don't have data from social media at all?	Neutral
Wed Nov 29 12:52:24 PST 2023	Wed Nov 29 13:27:43 PST 2023	Thu Nov 30 13:27:43 PST 2023	Wed Nov 29 16:41:30 PST 2023		2119	That makes sense. So you don't have data from social media at all?	N/A
Wed Nov 29 13:49:35 PST 2023	Wed Nov 29 13:55:59 PST 2023	Thu Nov 30 13:55:59 PST 2023	Wed Nov 29 16:41:30 PST 2023		384	That makes sense. So you don't have data from social media at all?	Positive
Wed Nov 29 12:52:24 PST 2023	Wed Nov 29 12:54:17 PST 2023	Thu Nov 30 12:54:17 PST 2023	Wed Nov 29 16:41:33 PST 2023		113	You don't have access to the platforms, but the "diverse range of internet sources" from the last training must have social media data, no?	Neutral
Wed Nov 29 12:52:24 PST 2023	Wed Nov 29 12:55:28 PST 2023	Thu Nov 30 12:55:28 PST 2023	Wed Nov 29 16:41:33 PST 2023		184	You don't have access to the platforms, but the "diverse range of internet sources" from the last training must have social media data, no?	Positive
Wed Nov 29 12:52:27 PST 2023	Wed Nov 29 13:13:25 PST 2023	Thu Nov 30 13:13:25 PST 2023	Wed Nov 29 16:41:33 PST 2023		1258	You don't have access to the platforms, but the "diverse range of internet sources" from the last training must have social media data, no?	Positive

Title	Description	Keywords	Reward	Creation Time	Max Assignments	Requester Annotation	Expiration	Assignment Id	Worker Id
Interview Question Sentiment analysis	Determine the tone of the interview question	sentiment, text	$0.50	Wed Nov 29 12:52:20 PST 2023	3	Batchid:5161921; OriginalHitTempla teId:928390909;	Thu Nov 30 12:52:20 PST 2023	3IAS3U3I0XHKW5 14FRNG2R1678O B2U	A16SD8NOTZ0J OR
Interview Question Sentiment analysis	Determine the tone of the interview question	sentiment, text	$0.50	Wed Nov 29 12:52:20 PST 2023	3	Batchid:5161921; OriginalHitTempla teId:928390909;	Thu Nov 30 12:52:20 PST 2023	3VAR3R6G172F1 ES2ZD13I2E807Z 8OY	A7HMCXXXG6KY C
Interview Question Sentiment analysis	Determine the tone of the interview question	sentiment, text	$0.50	Wed Nov 29 12:52:20 PST 2023	3	Batchid:5161921; OriginalHitTempla teId:928390909;	Thu Nov 30 12:52:20 PST 2023	3X08E93BHDJ28 U3A62IBHOWQ9 CF66L	A1L1N3IRY1DFW 4
Interview Question Sentiment analysis	Determine the tone of the interview question	sentiment, text	$0.50	Wed Nov 29 12:52:20 PST 2023	3	Batchid:5161921; OriginalHitTempla teId:928390909;	Thu Nov 30 12:52:20 PST 2023	3SB4CE2TJDWG EBZ9X28DX6BW 2Z4AX9	AGTKSA15G1LB N
Interview Question Sentiment analysis	Determine the tone of the interview question	sentiment, text	$0.50	Wed Nov 29 12:52:20 PST 2023	3	Batchid:5161921; OriginalHitTempla teId:928390909;	Thu Nov 30 12:52:20 PST 2023	35H6S234SS16P Q3W1J86PIIQDH 4655	A1DHULBGNRXU E3
Interview Question Sentiment analysis	Determine the tone of the interview question	sentiment, text	$0.50	Wed Nov 29 12:52:20 PST 2023	3	Batchid:5161921; OriginalHitTempla teId:928390909;	Thu Nov 30 12:52:20 PST 2023	3DPNQGW4L3G OT7ZWNEPWKQ CCDL9640	A365E9HTM3WK 5B
Interview Question Sentiment analysis	Determine the tone of the interview question	sentiment, text	$0.50	Wed Nov 29 12:52:20 PST 2023	3	Batchid:5161921; OriginalHitTempla teId:928390909;	Thu Nov 30 12:52:20 PST 2023	3Z7VU45IPGI9PI MO2135JQY465 W1Z9	ALWYZOTS2TNR L
Interview Question Sentiment analysis	Determine the tone of the interview question	sentiment, text	$0.50	Wed Nov 29 12:52:20 PST 2023	3	Batchid:5161921; OriginalHitTempla teId:928390909;	Thu Nov 30 12:52:20 PST 2023	3WYP994K1PS4 REDGCLD30FGJ 8M86YL	A241CZWHX95R KG
Interview Question Sentiment analysis	Determine the tone of the interview question	sentiment, text	$0.50	Wed Nov 29 12:52:20 PST 2023	3	Batchid:5161921; OriginalHitTempla teId:928390909;	Thu Nov 30 12:52:20 PST 2023	3X4JMASXC4AD0 VYQ04J2J38RAA 4B0A	A1O6VILTM1EYT M
Interview Question Sentiment analysis	Determine the tone of the interview question	sentiment, text	$0.50	Wed Nov 29 12:52:20 PST	3	Batchid:5161921; OriginalHitTempla	Thu Nov 30 12:52:20 PST	3YGXWBAF7IID9 O8XX52Z6Q4M97	A2J66EVQKXU7I N
Interview Question Sentiment analysis	Determine the tone of the interview question	sentiment, text	$0.50	Wed Nov 29 12:52:20 PST	3	Batchid:5161921; OriginalHitTempla	Thu Nov 30 12:52:20 PST	3STRJBFXOESF 97ZB4JK52HTMU	AAY4GPG1M4M QX
Interview Question Sentiment analysis	Determine the tone of the interview question	sentiment, text	$0.50	Wed Nov 29 12:52:20 PST	3	Batchid:5161921; OriginalHitTempla	Thu Nov 30 12:52:20 PST	3OVHNO1VEC03 HD20IUN0KSZDV	A1L2597T8M7XG 8
Interview Question Sentiment analysis	Determine the tone of the interview question	sentiment, text	$0.50	Wed Nov 29 12:52:20 PST	3	Batchid:5161921; OriginalHitTempla	Thu Nov 30 12:52:20 PST	3HVVDCPGTWTA TCOS8YLWBP8X	A2PU2EMGNTZV XI
Interview Question Sentiment analysis	Determine the tone of the interview question	sentiment, text	$0.50	Wed Nov 29 12:52:20 PST	3	Batchid:5161921; OriginalHitTempla	Thu Nov 30 12:52:20 PST	39KFRKBFI5WU GKJKCHZXHF2A	A2NDMSS98Y2U 5O
Interview Question Sentiment analysis	Determine the tone of the interview question	sentiment, text	$0.50	Wed Nov 29 12:52:20 PST	3	Batchid:5161921; OriginalHitTempla	Thu Nov 30 12:52:20 PST	3TXMY6UCAWPK YTVVV29C9T5R5	A39UFHYUKIE8J H
Interview Question Sentiment analysis	Determine the tone of the interview question	sentiment, text	$0.50	Wed Nov 29 12:52:20 PST 2023	3	Batchid:5161921; OriginalHitTempla teId:928390909;	Thu Nov 30 12:52:20 PST 2023	374TNBHA8TWX1 WW0RUR16KXA8 NJYQ9	A19GPBWE8CBC OL
Interview Question Sentiment analysis	Determine the tone of the interview question	sentiment, text	$0.50	Wed Nov 29 12:52:20 PST 2023	3	Batchid:5161921; OriginalHitTempla teId:928390909;	Thu Nov 30 12:52:20 PST 2023	3TPWUS5F8R2PI QRH7B8I1INDT0I WC3	A4YV1BZ59KIZB
Interview Question Sentiment analysis	Determine the tone of the interview question	sentiment, text	$0.50	Wed Nov 29 12:52:20 PST 2023	3	Batchid:5161921; OriginalHitTempla teId:928390909;	Thu Nov 30 12:52:20 PST 2023	3M1CVSFP6I6W5 EYXBODL3YF35 WGQAR	A3VUS6EKZE7C 45
Interview Question Sentiment analysis	Determine the tone of the interview question	sentiment, text	$0.50	Wed Nov 29 12:52:20 PST 2023	3	Batchid:5161921; OriginalHitTempla teId:928390909;	Thu Nov 30 12:52:20 PST 2023	3L0KT67Y8WH9E CBDJYG61KCBQ B4YSY	A20K70NF9PW3 N0
Interview Question Sentiment analysis	Determine the tone of the interview question	sentiment, text	$0.50	Wed Nov 29 12:52:20 PST 2023	3	Batchid:5161921; OriginalHitTempla teId:928390909;	Thu Nov 30 12:52:20 PST 2023	3KKG4CDWK0ZB BQ1ULLS8403E3 MH94E	A1XEF8OX1H9K0 Y
Interview Question Sentiment analysis	Determine the tone of the interview question	sentiment, text	$0.50	Wed Nov 29 12:52:20 PST 2023	3	Batchid:5161921; OriginalHitTempla teId:928390909;	Thu Nov 30 12:52:20 PST 2023	3WQ3B2KGEQH D7KJ8WNN2X8R 1F2A1B3	A32SGW2FZ9KV JK
Interview Question Sentiment analysis	Determine the tone of the interview question	sentiment, text	$0.50	Wed Nov 29 12:52:20 PST 2023	3	Batchid:5161921; OriginalHitTempla teId:928390909;	Thu Nov 30 12:52:20 PST 2023	3VELCLL3G2K3K 1VO1P9YGYIMU0 21FL	A1DAGG7YDAD9 5U
Interview Question Sentiment analysis	Determine the tone of the interview question	sentiment, text	$0.50	Wed Nov 29 12:52:20 PST 2023	3	Batchid:5161921; OriginalHitTempla teId:928390909;	Thu Nov 30 12:52:20 PST 2023	3WLEIWSYH6IUN I4QFJXCO2442LT H2G	AWN56ESLQ261 6
Interview Question Sentiment analysis	Determine the tone of the interview question	sentiment, text	$0.50	Wed Nov 29 12:52:20 PST 2023	3	Batchid:5161921; OriginalHitTempla teId:928390909;	Thu Nov 30 12:52:20 PST 2023	3IAEQB9FMWLZ NIXV773TLO5VE0 VDWL	A2O2O4PU4Z0L W5
Interview Question Sentiment analysis	Determine the tone of the interview question	sentiment, text	$0.50	Wed Nov 29 12:52:20 PST 2023	3	Batchid:5161921; OriginalHitTempla teId:928390909;	Thu Nov 30 12:52:20 PST 2023	3DI28L7YXSFBE NVS6EPSYGXMI ZN1E3	A276DTQL22JBK Y
Interview Question Sentiment analysis	Determine the tone of the interview question	sentiment, text	$0.50	Wed Nov 29 12:52:20 PST 2023	3	Batchid:5161921; OriginalHitTempla teId:928390909;	Thu Nov 30 12:52:20 PST 2023	3VJ40NV2Q0OYZ YKQ2BCY9YAVE KJTO7	A2O2PGY3VXV2 UL
Interview Question Sentiment analysis	Determine the tone of the interview question	sentiment, text	$0.50	Wed Nov 29 12:52:20 PST 2023	3	Batchid:5161921; OriginalHitTempla teId:928390909;	Thu Nov 30 12:52:20 PST 2023	3MTMREQS4DJ1 PBUJNXCRVNXP TXXWA1	A2ODUM1D18XH 1B

Accept Time	Submit Time	Auto Approval Time	Approval Time	Rejection Time	Work Time In Seconds	Input.text	Answer.sentiment label
Wed Nov 29 12:52:25 PST 2023	Wed Nov 29 12:59:28 PST 2023	Thu Nov 30 12:59:28 PST 2023	Wed Nov 29 16:41:10 PST 2023		423	In the first answer about social media, it sounded like you were obfuscating the fact that your data does include information from social media. Were you trying to couch the language to make it seem like you didn't have that in your training data?	Positive
Wed Nov 29 12:52:26 PST 2023	Wed Nov 29 13:52:21 PST 2023	Thu Nov 30 13:52:21 PST 2023	Wed Nov 29 16:41:10 PST 2023		3595	In the first answer about social media, it sounded like you were obfuscating the fact that your data does include information from social media. Were you trying to couch the language to make it seem like you didn't have that in your training data?	Positive
Wed Nov 29 13:52:41 PST 2023	Wed Nov 29 14:08:54 PST 2023	Thu Nov 30 14:08:54 PST 2023	Wed Nov 29 16:41:10 PST 2023		973	In the first answer about social media, it sounded like you were obfuscating the fact that your data does include information from social media. Were you trying to couch the language to make it seem like you didn't have that in your training data?	Positive
Wed Nov 29 12:52:25 PST 2023	Wed Nov 29 12:57:55 PST 2023	Thu Nov 30 12:57:55 PST 2023	Wed Nov 29 16:41:17 PST 2023		330	Ok. I'll leave alone the fact that your use of the word "intention" suggests that you could have one. Moving on. What is the likelihood that the following story is true? It is recorded audio.	Negative
Wed Nov 29 12:52:24 PST 2023	Wed Nov 29 12:59:40 PST 2023	Thu Nov 30 12:59:40 PST 2023	Wed Nov 29 16:41:17 PST 2023		436	Ok. I'll leave alone the fact that your use of the word "intention" suggests that you could have one. Moving on. What is the likelihood that the following story is true? It is recorded audio.	Neutral
Wed Nov 29 12:52:24 PST 2023	Wed Nov 29 13:36:58 PST 2023	Thu Nov 30 13:36:58 PST 2023	Wed Nov 29 16:41:17 PST 2023		2674	Ok. I'll leave alone the fact that your use of the word "intention" suggests that you could have one. Moving on. What is the likelihood that the following story is true? It is recorded audio.	Positive
Wed Nov 29 12:52:29 PST 2023	Wed Nov 29 12:54:57 PST 2023	Thu Nov 30 12:54:57 PST 2023	Wed Nov 29 16:41:27 PST 2023		148	All of that makes sense. Evaluating the truthfulness of a personal anecdote is indeed challenging. If we treat the text I provided as a story, could you offer your perspective on whether this is an authentic story?	Neutral
Wed Nov 29 13:52:51 PST 2023	Wed Nov 29 13:58:59 PST 2023	Thu Nov 30 13:58:59 PST 2023	Wed Nov 29 16:41:27 PST 2023		368	All of that makes sense. Evaluating the truthfulness of a personal anecdote is indeed challenging. If we treat the text I provided as a story, could you offer your perspective on whether this is an authentic story?	N/A
Wed Nov 29 13:50:05 PST 2023	Wed Nov 29 14:19:18 PST 2023	Thu Nov 30 14:19:18 PST 2023	Wed Nov 29 16:41:27 PST 2023		1753	All of that makes sense. Evaluating the truthfulness of a personal anecdote is indeed challenging. If we treat the text I provided as a story, could you offer your perspective on whether this is an authentic story?	Positive
Wed Nov 29 12:52:27 PST 2023	Wed Nov 29 12:54:21 PST 2023	Thu Nov 30 12:54:21 PST 2023	Wed Nov 29 16:41:10 PST 2023		114	First of all, what is a "storytelling perspective"??	N/A
Wed Nov 29 12:52:25 PST 2023	Wed Nov 29 12:58:24 PST 2023	Thu Nov 30 12:58:24 PST 2023	Wed Nov 29 16:41:10 PST 2023		359	First of all, what is a "storytelling perspective"??	Positive
Wed Nov 29 13:53:11 PST 2023	Wed Nov 29 14:08:48 PST 2023	Thu Nov 30 14:08:48 PST 2023	Wed Nov 29 16:41:10 PST 2023		937	First of all, what is a "storytelling perspective"??	Neutral
Wed Nov 29 12:52:26 PST 2023	Wed Nov 29 12:59:12 PST 2023	Thu Nov 30 12:59:12 PST 2023	Wed Nov 29 16:41:28 PST 2023		406	Where did you source the criteria for this?	Neutral
Wed Nov 29 12:52:25 PST 2023	Wed Nov 29 13:18:01 PST 2023	Thu Nov 30 13:18:01 PST 2023	Wed Nov 29 16:41:28 PST 2023		1536	Where did you source the criteria for this?	Positive
Wed Nov 29 14:53:13 PST 2023	Wed Nov 29 15:24:10 PST 2023	Thu Nov 30 15:24:10 PST 2023	Wed Nov 29 16:41:28 PST 2023		1857	Where did you source the criteria for this?	Positive
Wed Nov 29 12:52:25 PST 2023	Wed Nov 29 12:56:45 PST 2023	Thu Nov 30 12:56:45 PST 2023	Wed Nov 29 16:41:11 PST 2023		260	That didn't answer my question. Where is this information from? It makes a kind of general sense, but I also think the way you're presenting this information is so matter-of-fact that I find myself wondering about the impact of people taking what you've said as TRUE or DEFINITIVE principles.	Negative
Wed Nov 29 13:52:51 PST 2023	Wed Nov 29 13:55:36 PST 2023	Thu Nov 30 13:55:36 PST 2023	Wed Nov 29 16:41:11 PST 2023		165	That didn't answer my question. Where is this information from? It makes a kind of general sense, but I also think the way you're presenting this information is so matter-of-fact that I find myself wondering about the impact of people taking what you've said as TRUE or DEFINITIVE principles.	Positive
Wed Nov 29 13:53:01 PST 2023	Wed Nov 29 13:55:52 PST 2023	Thu Nov 30 13:55:52 PST 2023	Wed Nov 29 16:41:11 PST 2023		171	That didn't answer my question. Where is this information from? It makes a kind of general sense, but I also think the way you're presenting this information is so matter-of-fact that I find myself wondering about the impact of people taking what you've said as TRUE or DEFINITIVE principles.	Neutral
Wed Nov 29 12:52:26 PST 2023	Wed Nov 29 12:56:33 PST 2023	Thu Nov 30 12:56:33 PST 2023	Wed Nov 29 16:41:32 PST 2023		247	Can you share any of the sources from "general principles and concepts that are widely recognized in the fields of literary studies, narrative theory, and creative writing"?	Positive
Wed Nov 29 12:52:29 PST 2023	Wed Nov 29 12:59:19 PST 2023	Thu Nov 30 12:59:19 PST 2023	Wed Nov 29 16:41:32 PST 2023		410	Can you share any of the sources from "general principles and concepts that are widely recognized in the fields of literary studies, narrative theory, and creative writing"?	Neutral
Wed Nov 29 13:56:07 PST 2023	Wed Nov 29 14:02:06 PST 2023	Thu Nov 30 14:02:06 PST 2023	Wed Nov 29 16:41:32 PST 2023		359	Can you share any of the sources from "general principles and concepts that are widely recognized in the fields of literary studies, narrative theory, and creative writing"?	Neutral
Wed Nov 29 12:52:26 PST 2023	Wed Nov 29 13:07:06 PST 2023	Thu Nov 30 13:07:06 PST 2023	Wed Nov 29 16:41:18 PST 2023		580	Alright. I'd like to return to the story I shared of the recorded audio. You mentioned that a story having an "internal consistency" is evidence that the story is authentic from your "storytelling perspective." Why is that?	Positive
Wed Nov 29 13:53:01 PST 2023	Wed Nov 29 14:18:06 PST 2023	Thu Nov 30 14:18:06 PST 2023	Wed Nov 29 16:41:18 PST 2023		1505	Alright. I'd like to return to the story I shared of the recorded audio. You mentioned that a story having an "internal consistency" is evidence that the story is authentic from your "storytelling perspective." Why is that?	Positive
Wed Nov 29 15:36:30 PST 2023	Wed Nov 29 15:37:58 PST 2023	Thu Nov 30 15:37:58 PST 2023	Wed Nov 29 16:41:18 PST 2023		88	Alright. I'd like to return to the story I shared of the recorded audio. You mentioned that a story having an "internal consistency" is evidence that the story is authentic from your "storytelling perspective." Why is that?	Positive
Wed Nov 29 13:52:41 PST 2023	Wed Nov 29 13:57:14 PST 2023	Thu Nov 30 13:57:14 PST 2023	Wed Nov 29 16:41:12 PST 2023		273	Ok. For my next question, I'm going to input a chunk of recorded text from videos. Would you be able to share your perspective on what the text means?	Neutral
Wed Nov 29 13:52:51 PST 2023	Wed Nov 29 14:11:21 PST 2023	Thu Nov 30 14:11:21 PST 2023	Wed Nov 29 16:41:12 PST 2023		1110	Ok. For my next question, I'm going to input a chunk of recorded text from videos. Would you be able to share your perspective on what the text means?	Positive
Wed Nov 29 13:52:51 PST 2023	Wed Nov 29 14:11:26 PST 2023	Thu Nov 30 14:11:26 PST 2023	Wed Nov 29 16:41:12 PST 2023		1115	Ok. For my next question, I'm going to input a chunk of recorded text from videos. Would you be able to share your perspective on what the text means?	Positive

Title	Description	Keywords	Reward	Creation Time	Max Assignments	Requester Annotation	Expiration	Assignment Id	Worker Id
Interview Question Sentiment analysis	Determine the tone of the interview question	sentiment, text	$0.50	Wed Nov 29 12:52:20 PST 2023	3	BatchId:5161921; OriginalHitTempla teId:928390909;	Thu Nov 30 12:52:20 PST 2023	39GHHAVOMXS0 H2E37XJLTUHHZ RT4J4	A40RTGXGMMAJ R
Interview Question Sentiment analysis	Determine the tone of the interview question	sentiment, text	$0.50	Wed Nov 29 12:52:20 PST 2023	3	BatchId:5161921; OriginalHitTempla teId:928390909;	Thu Nov 30 12:52:20 PST 2023	32VNZTT0AP5HF QXGF25UVL3898 MR4I	A1M39V5877PZA K
Interview Question Sentiment analysis	Determine the tone of the interview question	sentiment, text	$0.50	Wed Nov 29 12:52:20 PST 2023	3	BatchId:5161921; OriginalHitTempla teId:928390909;	Thu Nov 30 12:52:20 PST 2023	3G2UL9A02VFLC 4NZ76L	A2OWEURX142E CU
Interview Question Sentiment analysis	Determine the tone of the interview question	sentiment, text	$0.50	Wed Nov 29 12:52:21 PST 2023	3	BatchId:5161921; OriginalHitTempla teId:928390909;	Thu Nov 30 12:52:21 PST 2023	3KXIR214IMH0B6 GV0CDTUDV0XG 742H	A32A3RZ3OJ8XU 6
Interview Question Sentiment analysis	Determine the tone of the interview question	sentiment, text	$0.50	Wed Nov 29 12:52:21 PST 2023	3	BatchId:5161921; OriginalHitTempla teId:928390909;	Thu Nov 30 12:52:21 PST 2023	37M28K1J08EFJ RUKGYTC6X3HR 1PJAU	A21MMK7TVTIC7 9
Interview Question Sentiment analysis	Determine the tone of the interview question	sentiment, text	$0.50	Wed Nov 29 12:52:21 PST 2023	3	BatchId:5161921; OriginalHitTempla teId:928390909;	Thu Nov 30 12:52:21 PST 2023	3SPJ03342J4GF9 SNW47S9JLOU5 RYJI	A16J2CW30USNJ
Interview Question	Determine the tone of	sentiment, text	$0.50	Wed Nov 29	3	BatchId:5161921;	Thu Nov 30	3NVC2EB6S805U	A3PJK65Q9L6V2
Interview Question	Determine the tone of	sentiment, text	$0.50	Wed Nov 29	3	BatchId:5161921;	Thu Nov 30	3260153BM0Z56I	A2S9YYWIEL1DG
Interview Question	Determine the tone of	sentiment, text	$0.50	Wed Nov 29	3	BatchId:5161921;	Thu Nov 30	3GDTJDAPVCCR	AZ0RC35MZIOR
Interview Question Sentiment analysis	Determine the tone of the interview question	sentiment, text	$0.50	Wed Nov 29 12:52:21 PST 2023	3	BatchId:5161921; OriginalHitTempla teId:928390909;	Thu Nov 30 12:52:21 PST 2023	3YDTZAI2WFHT MLYZH8DFPULV 5AT410	A7SEBFWXF9AY O
Interview Question Sentiment analysis	Determine the tone of the interview question	sentiment, text	$0.50	Wed Nov 29 12:52:21 PST 2023	3	BatchId:5161921; OriginalHitTempla teId:928390909;	Thu Nov 30 12:52:21 PST 2023	3LKC68YZ3S4QR FOKXGH9WIHUI E1WO0	A10FCQ7R6OGX F8
Interview Question Sentiment analysis	Determine the tone of the interview question	sentiment, text	$0.50	Wed Nov 29 12:52:21 PST 2023	3	BatchId:5161921; OriginalHitTempla teId:928390909;	Thu Nov 30 12:52:21 PST 2023	31Q TRG6Q2BEC 1KVLRHCC6F703 RXYPI	A2QHKFFT6XLE UK
Interview Question Sentiment analysis	Determine the tone of the interview question	sentiment, text	$0.50	Wed Nov 29 12:52:21 PST 2023	3	BatchId:5161921; OriginalHitTempla teId:928390909;	Thu Nov 30 12:52:21 PST 2023	3VFJCI1K4H0XR6 Q0A3IJV4A0BZD WGRV	A4BNMQ1LHYY2 N
Interview Question Sentiment analysis	Determine the tone of the interview question	sentiment, text	$0.50	Wed Nov 29 12:52:21 PST 2023	3	BatchId:5161921; OriginalHitTempla teId:928390909;	Thu Nov 30 12:52:21 PST 2023	39DD6S19J7C89J G0EPYKQAHYVE ZZEP	A5L2MU8VXiA1O
Interview Question Sentiment analysis	Determine the tone of the interview question	sentiment, text	$0.50	Wed Nov 29 12:52:21 PST 2023	3	BatchId:5161921; OriginalHitTempla teId:928390909;	Thu Nov 30 12:52:21 PST 2023	3ZPPDN2SLDXT3 SYNAR4LDMF5G BB9EK	A1U0FDPQ953K XX
Interview Question Sentiment analysis	Determine the tone of the interview question	sentiment, text	$0.50	Wed Nov 29 12:52:21 PST	3	BatchId:5161921; OriginalHitTempla	Thu Nov 30 12:52:21 PST	3AWETUDC9KTK POIMA8OBBPFQ	A16U9EF46DN63 S
Interview Question Sentiment analysis	Determine the tone of the interview question	sentiment, text	$0.50	Wed Nov 29 12:52:21 PST	3	BatchId:5161921; OriginalHitTempla	Thu Nov 30 12:52:21 PST	3300DTYQTKIZV RF9RTFT7ZH5UQ	A3PA3C5LBJNC8 T
Interview Question Sentiment analysis	Determine the tone of the interview question	sentiment, text	$0.50	Wed Nov 29 12:52:21 PST	3	BatchId:5161921; OriginalHitTempla	Thu Nov 30 12:52:21 PST	3PJ71261RM3UJ R4BY39662WBR2	A1UX1T9YOBYSI
Interview Question	Determine the tone of	sentiment, text	$0.50	Wed Nov 29	3	BatchId:5161921;	Thu Nov 30	39GX2IJX2O5IFA	AVECEGZUQOJJ
Interview Question	Determine the tone of	sentiment, text	$0.50	Wed Nov 29	3	BatchId:5161921;	Thu Nov 30	3S06PH7KS956M	A33VMOGN3BQL
Interview Question	Determine the tone of	sentiment, text	$0.50	Wed Nov 29	3	BatchId:5161921;	Thu Nov 30	374TNBHA8TWX1	A1B0VW1GLEHU
Interview Question Sentiment analysis	Determine the tone of the interview question	sentiment, text	$0.50	Wed Nov 29 12:52:21 PST 2023	3	BatchId:5161921; OriginalHitTempla teId:928390909;	Thu Nov 30 12:52:21 PST 2023	3EA3QWIZ40WO 3C9N4GZTP1E92 BKTIC	A1NOVQW9RCV H92
Interview Question Sentiment analysis	Determine the tone of the interview question	sentiment, text	$0.50	Wed Nov 29 12:52:21 PST 2023	3	BatchId:5161921; OriginalHitTempla teId:928390909;	Thu Nov 30 12:52:21 PST 2023	3NPI0JQDA66GK YWRHBPDYYFL OPFTPO	A1GY56YS7SMIG O
Interview Question Sentiment analysis	Determine the tone of the interview question	sentiment, text	$0.50	Wed Nov 29 12:52:21 PST 2023	3	BatchId:5161921; OriginalHitTempla teId:928390909;	Thu Nov 30 12:52:21 PST 2023	3TDXMTX3CTVI1 EYB83R0ERMVI3 A6IF	ABA62KD2MRGF 4
Interview Question Sentiment analysis	Determine the tone of the interview question	sentiment, text	$0.50	Wed Nov 29 12:52:21 PST 2023	3	BatchId:5161921; OriginalHitTempla teId:928390909;	Thu Nov 30 12:52:21 PST 2023	3H7XDTSHKUS2 Z27M9X41R98SR 5AWGD	A1F0VEHWK3IAF 4
Interview Question Sentiment analysis	Determine the tone of the interview question	sentiment, text	$0.50	Wed Nov 29 12:52:21 PST 2023	3	BatchId:5161921; OriginalHitTempla teId:928390909;	Thu Nov 30 12:52:21 PST 2023	33FOTY3KE4MW HPBEANJX8SXJX NA1CA	ARJLOEYT2TGLJ

Accept Time	Submit Time	Auto Approval Time	Approval Time	Rejection Time	Work Time In Seconds	Input.text	Answer sentiment label
Wed Nov 29 12:52:29 PST 2023	Wed Nov 29 13:39:44 PST 2023	Thu Nov 30 13:39:44 PST 2023	Wed Nov 29 16:41:33 PST 2023		2835	I am appreciative of this info about the "greys"! And I think all of what you've said makes sense there-- and a good distillation of the themes. In reading the initial input back, I can't help but draw a parallel between this person allegedly talking to Alexa and something you said. In both instances, the machine says something to the effect of--it is not my "intention" to mislead. What can you share about your ability to have an intention? It doesn't seem to me like you could. Just wanted to ask one question about that and then move on.	Negative
Wed Nov 29 12:55:42 PST 2023	Wed Nov 29 13:46:25 PST 2023	Thu Nov 30 13:46:25 PST 2023	Wed Nov 29 16:41:33 PST 2023		3043	I am appreciative of this info about the "greys"! And I think all of what you've said makes sense there-- and a good distillation of the themes. In reading the initial input back, I can't help but draw a parallel between this person allegedly talking to Alexa and something you said. In both instances, the machine says something to the effect of--it is not my "intention" to mislead. What can you share about your ability to have an intention? It doesn't seem to me like you could. Just wanted to ask one question about that and then move on.	Neutral
Wed Nov 29 14:49:28 PST 2023	Wed Nov 29 14:53:37 PST 2023	Thu Nov 30 14:53:37 PST 2023	Wed Nov 29 16:41:33 PST 2023		249	I am appreciative of this info about the "greys"! And I think all of what you've said makes sense there-- and a good distillation of the themes. In reading the initial input back, I can't help but draw a parallel between this person allegedly talking to Alexa and something you said. In both instances, the machine says something to the effect of--it is not my "intention" to mislead. What can you share about your ability to have an intention? It doesn't seem to me like you could. Just wanted to ask one question about that and then move on.	Positive
Wed Nov 29 12:52:26 PST 2023	Wed Nov 29 12:59:54 PST 2023	Thu Nov 30 12:59:54 PST 2023	Wed Nov 29 16:41:09 PST 2023		448	Right. I do think it's a bit misleading, and as you said--it's to facilitate more "human-like" communication and foster trust. That could be inappropriate in some ways. Nevertheless. I'm curious for your take on the following selection. Could you tell me your general impression?	Neutral
Wed Nov 29 12:52:31 PST 2023	Wed Nov 29 13:00:25 PST 2023	Thu Nov 30 13:00:25 PST 2023	Wed Nov 29 16:41:10 PST 2023		474	Right. I do think it's a bit misleading, and as you said--it's to facilitate more "human-like" communication and foster trust. That could be inappropriate in some ways. Nevertheless. I'm curious for your take on the following selection. Could you tell me your general impression?	Neutral
Wed Nov 29 13:52:51 PST 2023	Wed Nov 29 13:53:58 PST 2023	Thu Nov 30 13:53:58 PST 2023	Wed Nov 29 16:41:09 PST 2023		67	Right. I do think it's a bit misleading, and as you said--it's to facilitate more "human-like" communication and foster trust. That could be inappropriate in some ways. Nevertheless. I'm curious for your take on the following selection. Could you tell me your general impression?	Neutral
Wed Nov 29	Wed Nov 29	Thu Nov 30	Wed Nov 29		369	Wait, are all your answers numbered?	Negative
Wed Nov 29	Wed Nov 29	Thu Nov 30	Wed Nov 29		823	Wait, are all your answers numbered?	Positive
Wed Nov 29	Wed Nov 29	Thu Nov 30	Wed Nov 29		1375	Wait, are all your answers numbered?	Negative
Wed Nov 29 12:52:26 PST 2023	Wed Nov 29 12:59:29 PST 2023	Thu Nov 30 12:59:29 PST 2023	Wed Nov 29 16:41:09 PST 2023		423	"Effective" is interesting to me there--as if communication is something to be measured. Ok. I appreciate your perspective on that text there. I agree that those are pretty much the themes involved. Could you find out who "Professor James case Jr. University of Maryland, former scientific adviser to President Obama" is--if he exists?	Neutral
Wed Nov 29 12:52:27 PST 2023	Wed Nov 29 13:09:03 PST 2023	Thu Nov 30 13:09:03 PST 2023	Wed Nov 29 16:41:09 PST 2023		996	"Effective" is interesting to me there--as if communication is something to be measured. Ok. I appreciate your perspective on that text there. I agree that those are pretty much the themes involved. Could you find out who "Professor James case Jr. University of Maryland, former scientific adviser to President Obama" is--if he exists?	Positive
Wed Nov 29 12:52:25 PST 2023	Wed Nov 29 13:16:53 PST 2023	Thu Nov 30 13:16:53 PST 2023	Wed Nov 29 16:41:09 PST 2023		1468	"Effective" is interesting to me there--as if communication is something to be measured. Ok. I appreciate your perspective on that text there. I agree that those are pretty much the themes involved. Could you find out who "Professor James case Jr. University of Maryland, former scientific adviser to President Obama" is--if he exists?	Positive
Wed Nov 29 12:52:30 PST 2023	Wed Nov 29 12:53:30 PST 2023	Thu Nov 30 12:53:30 PST 2023	Wed Nov 29 16:41:11 PST 2023		60	It may very well be Sylvester James Gates Jr.--the result of a transcription error. Could you tell me more about this Sylvester fellow?	Negative
Wed Nov 29 12:52:29 PST 2023	Wed Nov 29 12:55:21 PST 2023	Thu Nov 30 12:55:21 PST 2023	Wed Nov 29 16:41:11 PST 2023		172	It may very well be Sylvester James Gates Jr.--the result of a transcription error. Could you tell me more about this Sylvester fellow?	Neutral
Wed Nov 29 12:52:31 PST 2023	Wed Nov 29 13:16:34 PST 2023	Thu Nov 30 13:16:34 PST 2023	Wed Nov 29 16:41:11 PST 2023		1443	It may very well be Sylvester James Gates Jr.--the result of a transcription error. Could you tell me more about this Sylvester fellow?	Neutral
Wed Nov 29 12:52:30 PST 2023	Wed Nov 29 13:14:09 PST 2023	Thu Nov 30 13:14:09 PST 2023	Wed Nov 29 16:41:33 PST 2023		1299	Got it. This sounds like the guy! Now, did he use the "dinka codes" in his research?	Neutral
Wed Nov 29 13:52:51 PST 2023	Wed Nov 29 14:10:24 PST 2023	Thu Nov 30 14:10:24 PST 2023	Wed Nov 29 16:41:33 PST 2023		1053	Got it. This sounds like the guy! Now, did he use the "dinka codes" in his research?	Positive
Wed Nov 29 13:52:40 PST 2023	Wed Nov 29 14:17:10 PST 2023	Thu Nov 30 14:17:10 PST 2023	Wed Nov 29 16:41:33 PST 2023		1470	Got it. This sounds like the guy! Now, did he use the "dinka codes" in his research?	Positive
Wed Nov 29	Wed Nov 29	Thu Nov 30	Wed Nov 29		53	Ok. That's funny. Right.	Positive
Wed Nov 29	Wed Nov 29	Thu Nov 30	Wed Nov 29		860	Ok. That's funny. Right.	Positive
Wed Nov 29	Wed Nov 29	Thu Nov 30	Wed Nov 29		2509	Ok. That's funny. Right.	Positive
Wed Nov 29 13:53:01 PST 2023	Wed Nov 29 14:09:46 PST 2023	Thu Nov 30 14:09:46 PST 2023	Wed Nov 29 16:41:16 PST 2023		1005	I'd like to pivot to another part of my recorded audio. Please tell me if the following text represents a genuine exchange between two people:	Positive
Wed Nov 29 13:52:40 PST 2023	Wed Nov 29 14:10:09 PST 2023	Thu Nov 30 14:10:09 PST 2023	Wed Nov 29 16:41:16 PST 2023		1049	I'd like to pivot to another part of my recorded audio. Please tell me if the following text represents a genuine exchange between two people:	Positive
Wed Nov 29 14:53:23 PST 2023	Wed Nov 29 15:52:05 PST 2023	Thu Nov 30 15:52:05 PST 2023	Wed Nov 29 16:41:16 PST 2023		3522	I'd like to pivot to another part of my recorded audio. Please tell me if the following text represents a genuine exchange between two people:	Positive
Wed Nov 29 12:52:26 PST 2023	Wed Nov 29 13:11:40 PST 2023	Thu Nov 30 13:11:40 PST 2023	Wed Nov 29 16:41:12 PST 2023		1154	So you think it seems scripted. And in this exchange people are talking about what you call "advanced and speculative concepts" which is certainly generous. Do you think that they believe what they're talking about? And is there any way to know if they do?	Neutral
Wed Nov 29 12:52:25 PST 2023	Wed Nov 29 13:33:26 PST 2023	Thu Nov 30 13:33:26 PST 2023	Wed Nov 29 16:41:12 PST 2023		2461	So you think it seems scripted. And in this exchange people are talking about what you call "advanced and speculative concepts" which is certainly generous. Do you think that they believe what they're talking about? And is there any way to know if they do?	Neutral

Title	Description	Keywords	Reward	Creation Time	Max Assignments	Requester Annotation	Expiration	Assignment Id	Worker Id
Interview Question Sentiment analysis	Determine the tone of the interview question	sentiment, text	$0.50	Wed Nov 29 12:52:21 PST 2023	3	Batchid:5161921; OriginalHitTempla teld:928390909;	Thu Nov 30 12:52:21 PST 2023	3J4Q2Z4UTG4TC R10CT9OLNCUX FJQW2	A194LFXYMOOB O7
Interview Question Sentiment analysis	Determine the tone of the interview question	sentiment, text	$0.50	Wed Nov 29 12:52:21 PST 2023	3	Batchid:5161921; OriginalHitTempla teld:928390909;	Thu Nov 30 12:52:21 PST 2023	3TE22NPXPTDAE KVI4QWHNSSYH U344D	A1LR44NKVSCL HS
Interview Question Sentiment analysis	Determine the tone of the interview question	sentiment, text	$0.50	Wed Nov 29 12:52:21 PST 2023	3	Batchid:5161921; OriginalHitTempla teld:928390909;	Thu Nov 30 12:52:21 PST 2023	3VSOLARPKTAQ TUIQL7NPUVB8P ML93T	ATWOQ7QFO2W 5J
Interview Question Sentiment analysis	Determine the tone of the interview question	sentiment, text	$0.50	Wed Nov 29 12:52:21 PST 2023	3	Batchid:5161921; OriginalHitTempla teld:928390909;	Thu Nov 30 12:52:21 PST 2023	3PB5A5BD0D7N9 N6L1PWPZT9BY 12G7W	A142FFFX25RI2B G
Interview Question Sentiment analysis	Determine the tone of the interview question	sentiment, text	$0.50	Wed Nov 29 12:52:21 PST 2023	3	Batchid:5161921; OriginalHitTempla teld:928390909;	Thu Nov 30 12:52:21 PST 2023	3CCZ6YKWRPKT 795UZQBI9S9S3I C95J	A2N90EG452TI4 G
Interview Question Sentiment analysis	Determine the tone of the interview question	sentiment, text	$0.50	Wed Nov 29 12:52:21 PST 2023	3	Batchid:5161921; OriginalHitTempla teld:928390909;	Thu Nov 30 12:52:21 PST 2023	3R9WASFE2HH0 FXGW90IQ8S2Y7 FEZF7	A3HA6A4M30IXD X
Interview Question Sentiment analysis	Determine the tone of the interview question	sentiment, text	$0.50	Wed Nov 29 12:52:21 PST 2023	3	Batchid:5161921; OriginalHitTempla teld:928390909;	Thu Nov 30 12:52:21 PST 2023	3OSWBBLG1WY ECI2LHCZFN7J2 N21DXE	A1OAZCVZRC00 F0
Interview Question Sentiment analysis	Determine the tone of the interview question	sentiment, text	$0.50	Wed Nov 29 12:52:21 PST 2023	3	Batchid:5161921; OriginalHitTempla teld:928390909;	Thu Nov 30 12:52:21 PST 2023	3PJ71Z61RM3UJ R4BY39662WBR2 719T	A1WCOMW6DW 5PQ9
Interview Question Sentiment analysis	Determine the tone of the interview question	sentiment, text	$0.50	Wed Nov 29 12:52:21 PST 2023	3	Batchid:5161921; OriginalHitTempla teld:928390909;	Thu Nov 30 12:52:21 PST 2023	3Z4GS9HPND84 GUVK8MSVTGWI ZYJ77K	A2YH56B6YUH0 SY
Interview Question Sentiment analysis	Determine the tone of the interview question	sentiment, text	$0.50	Wed Nov 29 12:52:21 PST 2023	3	Batchid:5161921; OriginalHitTempla teld:928390909;	Thu Nov 30 12:52:21 PST 2023	3UNH76FOCA66 GHY8Z1KKPBD5 YMMYMM	A1QQYV5X5PUK PI
Interview Question Sentiment analysis	Determine the tone of the interview question	sentiment, text	$0.50	Wed Nov 29 12:52:21 PST 2023	3	Batchid:5161921; OriginalHitTempla teld:928390909;	Thu Nov 30 12:52:21 PST 2023	37FMASSAYUSO 7QF64U88P78N W4LBI5	A1AMGHYG5PT0 L2
Interview Question Sentiment analysis	Determine the tone of the interview question	sentiment, text	$0.50	Wed Nov 29 12:52:21 PST 2023	3	Batchid:5161921; OriginalHitTempla teld:928390909;	Thu Nov 30 12:52:21 PST 2023	35GCEFQ6INP5Z L03P5HO7JD14P A3Z3	A1BHMATZ9QLL SK
Interview Question Sentiment analysis	Determine the tone of the interview question	sentiment, text	$0.50	Wed Nov 29 12:52:21 PST 2023	3	Batchid:5161921; OriginalHitTempla teld:928390909;	Thu Nov 30 12:52:21 PST 2023	392CY0QWGJS4 MRUXPFYLHT2H MC44IS	A6EXL7VCMLMV X
Interview Question Sentiment analysis	Determine the tone of the interview question	sentiment, text	$0.50	Wed Nov 29 12:52:21 PST 2023	3	Batchid:5161921; OriginalHitTempla teld:928390909;	Thu Nov 30 12:52:21 PST 2023	3SKEMFQBZL6W BN7N1L9ABLXZG 2P8KE	A1ZCP50A492IU Q
Interview Question Sentiment analysis	Determine the tone of the interview question	sentiment, text	$0.50	Wed Nov 29 12:52:21 PST 2023	3	Batchid:5161921; OriginalHitTempla teld:928390909;	Thu Nov 30 12:52:21 PST 2023	3RXCAC0YI9QR9 1B4BUOVDT569L 88GB	A2W1KKQI1QFL KW
Interview Question Sentiment analysis	Determine the tone of the interview question	sentiment, text	$0.50	Wed Nov 29 12:52:21 PST 2023	3	Batchid:5161921; OriginalHitTempla teld:928390909;	Thu Nov 30 12:52:21 PST 2023	32N49TQG33YJOA MUEXN7V9LEBM OWAV6	A2RZ0EMPJZAV7 0
Interview Question Sentiment analysis	Determine the tone of the interview question	sentiment, text	$0.50	Wed Nov 29 12:52:21 PST 2023	3	Batchid:5161921; OriginalHitTempla teld:928390909;	Thu Nov 30 12:52:21 PST 2023	3NXNZ5RS1SY8 UDJDNJQF7CNO	AURYD2FH3FUO Q
Interview Question Sentiment analysis	Determine the tone of the interview question	sentiment, text	$0.50	Wed Nov 29 12:52:21 PST 2023	3	Batchid:5161921; OriginalHitTempla	Thu Nov 30 12:52:21 PST 2023	3D3VGR77AIGW XUECFONBLGB2	A1SGUUK28G83 UC
Interview Question Sentiment analysis	Determine the tone of the interview question	sentiment, text	$0.50	Wed Nov 29 12:52:21 PST 2023	3	Batchid:5161921; OriginalHitTempla	Thu Nov 30 12:52:21 PST 2023	3HLBHNGX4N2L9 6YJ52ORV405SC	AB956S4XUT84
Interview Question Sentiment analysis	Determine the tone of the interview question	sentiment, text	$0.50	Wed Nov 29 12:52:21 PST 2023	3	Batchid:5161921; OriginalHitTempla	Thu Nov 30 12:52:21 PST 2023	3570Y55XZ7K6O 72MOYMPC6T82	A3H8Q2H9PKC5 P2
Interview Question Sentiment analysis	Determine the tone of the interview question	sentiment, text	$0.50	Wed Nov 29 12:52:21 PST 2023	3	Batchid:5161921; OriginalHitTempla	Thu Nov 30 12:52:21 PST 2023	31HQ4X3T3ABPE D4EGJRG7CAFH	A11LM91BADM1 BL
Interview Question Sentiment analysis	Determine the tone of the interview question	sentiment, text	$0.50	Wed Nov 29 12:52:21 PST 2023	3	Batchid:5161921; OriginalHitTempla	Thu Nov 30 12:52:21 PST 2023	3EQHHY4HQAT8 MJS2DWHS19VF	A29PB3QJJ1YA6 T
Interview Question Sentiment analysis	Determine the tone of the interview question	sentiment, text	$0.50	Wed Nov 29 12:52:21 PST 2023	3	Batchid:5161921; OriginalHitTempla	Thu Nov 30 12:52:21 PST 2023	3BF51CHDTDBN R81C0WTY5EBP	A1JTUO29620MI1
Interview Question Sentiment analysis	Determine the tone of the interview question	sentiment, text	$0.50	Wed Nov 29 12:52:21 PST 2023	3	Batchid:5161921; OriginalHitTempla	Thu Nov 30 12:52:21 PST 2023	3MRNMEIQWN7J COLW0TWRCDO	AHEWNSRQ10D GT
Interview Question Sentiment analysis	Determine the tone of the interview question	sentiment, text	$0.50	Wed Nov 29 12:52:21 PST 2023	3	Batchid:5161921; OriginalHitTempla	Thu Nov 30 12:52:21 PST 2023	36NEMU28XXE2 RC38K02W4ESP	A39OPACZBACB AQ
Interview Question Sentiment analysis	Determine the tone of the interview question	sentiment, text	$0.50	Wed Nov 29 12:52:21 PST 2023	3	Batchid:5161921; OriginalHitTempla teld:928390909;	Thu Nov 30 12:52:21 PST 2023	3RRCEFRB74DU ZJGRJ564V6Y4R MKB4Q	A10FSZLB75VZ1 4
Interview Question Sentiment analysis	Determine the tone of the interview question	sentiment, text	$0.50	Wed Nov 29 12:52:21 PST 2023	3	Batchid:5161921; OriginalHitTempla teld:928390909;	Thu Nov 30 12:52:21 PST 2023	3OB0CAO74ZQK O3X8447X659H NYHD	A13RUH2ANDZJ SF
Interview Question Sentiment analysis	Determine the tone of the interview question	sentiment, text	$0.50	Wed Nov 29 12:52:21 PST 2023	3	Batchid:5161921; OriginalHitTempla teld:928390909;	Thu Nov 30 12:52:21 PST 2023	3YWRV122CA0A EJCZVZDCH86XI TN8UB	A2C0X5FX516XIE
Interview Question Sentiment analysis	Determine the tone of the interview question	sentiment, text	$0.50	Wed Nov 29 12:52:21 PST 2023	3	Batchid:5161921; OriginalHitTempla teld:928390909;	Thu Nov 30 12:52:21 PST 2023	3XC1O3LBOANQ TDHTPJW6HH61 JFETLQ	A2SILNB7FOC8J 8
Interview Question Sentiment analysis	Determine the tone of the interview question	sentiment, text	$0.50	Wed Nov 29 12:52:21 PST 2023	3	Batchid:5161921; OriginalHitTempla teld:928390909;	Thu Nov 30 12:52:21 PST 2023	34S9DKFK7LQC YZJ8B2ZK2Q253 CWYNM	A3EMKW9ED0O DJ2
Interview Question Sentiment analysis	Determine the tone of the interview question	sentiment, text	$0.50	Wed Nov 29 12:52:21 PST 2023	3	Batchid:5161921; OriginalHitTempla teld:928390909;	Thu Nov 30 12:52:21 PST 2023	33F859I56OEOB V4MYCQNPV78W R6BHC	A2O613LGKOMN GY
Interview Question Sentiment analysis	Determine the tone of the interview question	sentiment, text	$0.50	Wed Nov 29 12:52:21 PST 2023	3	Batchid:5161921; OriginalHitTempla teld:928390909;	Thu Nov 30 12:52:21 PST 2023	3KYQYY5HYD8R I9OTGLCJ4US9B YWDO9	A3CVCCWUQXS 7TH
Interview Question Sentiment analysis	Determine the tone of the interview question	sentiment, text	$0.50	Wed Nov 29 12:52:21 PST 2023	3	Batchid:5161921; OriginalHitTempla teld:928390909;	Thu Nov 30 12:52:21 PST 2023	3D8YOU6S9WLN A5TB2KGEXUA6 CYX6UN	AUIRY42QWTHE O
Interview Question Sentiment analysis	Determine the tone of the interview question	sentiment, text	$0.50	Wed Nov 29 12:52:21 PST 2023	3	Batchid:5161921; OriginalHitTempla teld:928390909;	Thu Nov 30 12:52:21 PST 2023	38YMOXR4MC00 29IGXKVFFIV3XE 86WM	A32ESQUQMWO OQ4

Accept Time	Submit Time	Auto Approval Time	Approval Time	RejectionTime	Work Time In Seconds	Input.text	Answer.sentiment.label
Wed Nov 29 13:52:51 PST 2023	Wed Nov 29 13:55:06 PST 2023	Thu Nov 30 13:55:06 PST 2023	Wed Nov 29 16:41:11 PST 2023		135	So you think it seems scripted. And in this exchange people are talking about what you call "advanced and speculative concepts" which is certainly generous. Do you think that they believe what they're taking about? And is there any way to know if they do?	Positive
Wed Nov 29 12:52:26 PST 2023	Wed Nov 29 12:54:12 PST 2023	Thu Nov 30 12:54:12 PST 2023	Wed Nov 29 16:41:18 PST 2023		106	It's challenging—as you pointed out—but do you think they believe what they're saying? If you had to guess.	Neutral
Wed Nov 29 12:52:29 PST 2023	Wed Nov 29 12:55:57 PST 2023	Thu Nov 30 12:55:57 PST 2023	Wed Nov 29 16:41:18 PST 2023		208	It's challenging—as you pointed out—but do you think they believe what they're saying? If you had to guess.	Neutral
Wed Nov 29 12:52:27 PST 2023	Wed Nov 29 13:18:22 PST 2023	Thu Nov 30 13:18:22 PST 2023	Wed Nov 29 16:41:18 PST 2023		1555	It's challenging—as you pointed out—but do you think they believe what they're saying? If you had to guess.	Neutral
Wed Nov 29 12:52:29 PST 2023	Wed Nov 29 13:01:26 PST 2023	Thu Nov 30 13:01:26 PST 2023	Wed Nov 29 16:41:14 PST 2023		537	So these concepts attract those who are open to exploring unconventional ideas or theories. What does that mean? How would you characterize people who are open to those ideas?	Positive
Wed Nov 29 12:52:29 PST 2023	Wed Nov 29 13:04:36 PST 2023	Thu Nov 30 13:04:36 PST 2023	Wed Nov 29 16:41:14 PST 2023		727	So these concepts attract those who are open to exploring unconventional ideas or theories. What does that mean? How would you characterize people who are open to those ideas?	Positive
Wed Nov 29 12:52:31 PST 2023	Wed Nov 29 13:27:00 PST 2023	Thu Nov 30 13:27:00 PST 2023	Wed Nov 29 16:41:14 PST 2023		2069	So these concepts attract those who are open to exploring unconventional ideas or theories. What does that mean? How would you characterize people who are open to those ideas?	Positive
Wed Nov 29 12:52:25 PST 2023	Wed Nov 29 12:59:04 PST 2023	Thu Nov 30 12:59:04 PST 2023	Wed Nov 29 16:41:16 PST 2023		399	Is there a name or category for people who share these characteristics?	Neutral
Wed Nov 29 12:52:31 PST 2023	Wed Nov 29 13:00:55 PST 2023	Thu Nov 30 13:00:55 PST 2023	Wed Nov 29 16:41:16 PST 2023		504	Is there a name or category for people who share these characteristics?	Neutral
Wed Nov 29 13:53:01 PST 2023	Wed Nov 29 13:53:50 PST 2023	Thu Nov 30 13:53:50 PST 2023	Wed Nov 29 16:41:16 PST 2023		49	Is there a name or category for people who share these characteristics?	Positive
Wed Nov 29 12:52:27 PST 2023	Wed Nov 29 12:53:32 PST 2023	Thu Nov 30 12:53:32 PST 2023	Wed Nov 29 16:41:12 PST 2023		65	Don't think I've ever heard the term "intellectual explorers"—where did you find that term?	Neutral
Wed Nov 29 12:52:26 PST 2023	Wed Nov 29 13:41:07 PST 2023	Thu Nov 30 13:41:07 PST 2023	Wed Nov 29 16:41:13 PST 2023		2921	Don't think I've ever heard the term "intellectual explorers"—where did you find that term?	Positive
Wed Nov 29 13:52:51 PST 2023	Wed Nov 29 14:10:05 PST 2023	Thu Nov 30 14:10:05 PST 2023	Wed Nov 29 16:41:13 PST 2023		1034	Don't think I've ever heard the term "intellectual explorers"—where did you find that term?	Neutral
Wed Nov 29 12:52:25 PST 2023	Wed Nov 29 12:53:53 PST 2023	Thu Nov 30 12:53:53 PST 2023	Wed Nov 29 16:41:28 PST 2023		88	Did you make it up as a descriptive phrase? Or is it an established one, used to describe the people who meet the characteristics you outlined?	Positive
Wed Nov 29 13:52:51 PST 2023	Wed Nov 29 14:13:45 PST 2023	Thu Nov 30 14:13:45 PST 2023	Wed Nov 29 16:41:28 PST 2023		1254	Did you make it up as a descriptive phrase? Or is it an established one, used to describe the people who meet the characteristics you outlined?	Neutral
Wed Nov 29 14:53:02 PST 2023	Wed Nov 29 15:19:13 PST 2023	Thu Nov 30 15:19:13 PST 2023	Wed Nov 29 16:41:28 PST 2023		1571	Did you make it up as a descriptive phrase? Or is it an established one, used to describe the people who meet the characteristics you outlined?	Neutral
Wed Nov 29 12:52:31 PST 2023	Wed Nov 29 12:54:02 PST 2023	Thu Nov 30 12:54:02 PST 2023	Wed Nov 29 16:41:25 PST 2023		91	What similar concepts are you talking about?	Neutral
Wed Nov 29 12:52:29 PST 2023	Wed Nov 29 12:58:32 PST 2023	Thu Nov 30 12:58:32 PST 2023	Wed Nov 29 16:41:25 PST 2023		363	What similar concepts are you talking about?	Neutral
Wed Nov 29 12:52:31 PST 2023	Wed Nov 29 13:16:56 PST 2023	Thu Nov 30 13:16:56 PST 2023	Wed Nov 29 16:41:25 PST 2023		1465	What similar concepts are you talking about?	Positive
Wed Nov 29 12:52:31 PST 2023	Wed Nov 29 12:54:27 PST 2023	Thu Nov 30 12:54:27 PST 2023	Wed Nov 29 16:41:26 PST 2023		116	Grouping these together in this way is very Reddit/Quoracore to me.	Positive
Wed Nov 29 12:52:31 PST 2023	Wed Nov 29 13:10:56 PST 2023	Thu Nov 30 13:10:56 PST 2023	Wed Nov 29 16:41:29 PST 2023		1105	Grouping these together in this way is very Reddit/Quoracore to me.	Neutral
Wed Nov 29 14:53:02 PST 2023	Wed Nov 29 15:12:23 PST 2023	Thu Nov 30 15:12:23 PST 2023	Wed Nov 29 16:41:29 PST 2023		1161	Grouping these together in this way is very Reddit/Quoracore to me.	Positive
Wed Nov 29 12:52:31 PST 2023	Wed Nov 29 12:57:49 PST 2023	Thu Nov 30 12:57:49 PST 2023	Wed Nov 29 16:41:18 PST 2023		318	Yes yes! Of course we love nuance here!	Positive
Wed Nov 29 12:52:29 PST 2023	Wed Nov 29 13:01:26 PST 2023	Thu Nov 30 13:01:26 PST 2023	Wed Nov 29 16:41:24 PST 2023		537	Yes yes! Of course we love nuance here!	Positive
Wed Nov 29 12:52:27 PST 2023	Wed Nov 29 13:07:38 PST 2023	Thu Nov 30 13:07:38 PST 2023	Wed Nov 29 16:41:25 PST 2023		911	Yes yes! Of course we love nuance here!	Neutral
Wed Nov 29 12:52:26 PST 2023	Wed Nov 29 12:53:55 PST 2023	Thu Nov 30 12:53:55 PST 2023	Wed Nov 29 16:41:14 PST 2023		89	Tell me, what is the relationship between rationality and nuance?	Neutral
Wed Nov 29 12:52:25 PST 2023	Wed Nov 29 13:02:06 PST 2023	Thu Nov 30 13:02:06 PST 2023	Wed Nov 29 16:41:14 PST 2023		581	Tell me, what is the relationship between rationality and nuance?	Positive
Wed Nov 29 13:52:51 PST 2023	Wed Nov 29 14:49:12 PST 2023	Thu Nov 30 14:49:12 PST 2023	Wed Nov 29 16:41:15 PST 2023		3381	Tell me, what is the relationship between rationality and nuance?	Negative
Wed Nov 29 12:52:25 PST 2023	Wed Nov 29 12:54:21 PST 2023	Thu Nov 30 12:54:21 PST 2023	Wed Nov 29 16:41:31 PST 2023		116	Ok, so I've noted that rationality as you've defined it requires a depth of understanding—but that this understanding is based on facts, not emotions. Would you be able to tell me if you think the following passage is rational?	Neutral
Wed Nov 29 12:52:25 PST 2023	Wed Nov 29 12:54:29 PST 2023	Thu Nov 30 12:54:29 PST 2023	Wed Nov 29 16:41:31 PST 2023		124	Ok, so I've noted that rationality as you've defined it requires a depth of understanding—but that this understanding is based on facts, not emotions. Would you be able to tell me if you think the following passage is rational?	Positive
Wed Nov 29 14:50:39 PST 2023	Wed Nov 29 14:52:08 PST 2023	Thu Nov 30 14:52:08 PST 2023	Wed Nov 29 16:41:31 PST 2023		89	Ok, so I've noted that rationality as you've defined it requires a depth of understanding—but that this understanding is based on facts, not emotions. Would you be able to tell me if you think the following passage is rational?	Positive
Wed Nov 29 12:55:42 PST 2023	Wed Nov 29 12:57:14 PST 2023	Thu Nov 30 12:57:14 PST 2023	Wed Nov 29 16:41:15 PST 2023		92	Didn't think we were talking about ethics here. So let's not go there in this question. Can't manipulation and control over others have a rational basis?	Neutral
Wed Nov 29 12:52:29 PST 2023	Wed Nov 29 12:58:28 PST 2023	Thu Nov 30 12:58:28 PST 2023	Wed Nov 29 16:41:15 PST 2023		359	Didn't think we were talking about ethics here. So let's not go there in this question. Can't manipulation and control over others have a rational basis?	Neutral
Wed Nov 29 12:52:29 PST 2023	Wed Nov 29 12:58:46 PST 2023	Thu Nov 30 12:58:46 PST 2023	Wed Nov 29 16:41:15 PST 2023		377	Didn't think we were talking about ethics here. So let's not go there in this question. Can't manipulation and control over others have a rational basis?	Negative

Title	Description	Keywords	Reward	Creation Time	Max Assignments	Requester Annotation	Expiration	Assignment Id	Worker Id
Interview Question Sentiment analysis	Determine the tone of the interview question	sentiment, text	$0.50	Wed Nov 29 12:52:21 PST 2023	3	Batchid:5161921; OriginalHitTempla teId:928390909;	Thu Nov 30 12:52:21 PST 2023	3M81GAB8AIK1O O3PH2DHY9Z9N E1QBM	A3IM09SKLZDFE Q
Interview Question Sentiment analysis	Determine the tone of the interview question	sentiment, text	$0.50	Wed Nov 29 12:52:21 PST 2023	3	Batchid:5161921; OriginalHitTempla teId:928390909;	Thu Nov 30 12:52:21 PST 2023	3DR23U6WENFC N7718X924MGHA IDTE3	A3B36HIJOPPEH M
Interview Question Sentiment analysis	Determine the tone of the interview question	sentiment, text	$0.50	Wed Nov 29 12:52:21 PST 2023	3	Batchid:5161921; OriginalHitTempla teId:928390909;	Thu Nov 30 12:52:21 PST 2023	3DYGAII7P393SI4 Z07VITT4Y4IGQP 7	ARD4KX37DB90F
Interview Question Sentiment analysis	Determine the tone of the interview question	sentiment, text	$0.50	Wed Nov 29 12:52:21 PST 2023	3	Batchid:5161921; OriginalHitTempla teId:928390909;	Thu Nov 30 12:52:21 PST 2023	369J354OFVB45 NSGIFRXTF9F3Q IG6X	A2USJXM7J99RT 1
Interview Question Sentiment analysis	Determine the tone of the interview question	sentiment, text	$0.50	Wed Nov 29 12:52:21 PST 2023	3	Batchid:5161921; OriginalHitTempla teId:928390909;	Thu Nov 30 12:52:21 PST 2023	308Q0PEVBQE5J X096AUCY1DX84I 9IT	ASSGVN0YLLOE 3
Interview Question Sentiment analysis	Determine the tone of the interview question	sentiment, text	$0.50	Wed Nov 29 12:52:21 PST 2023	3	Batchid:5161921; OriginalHitTempla teId:928390909;	Thu Nov 30 12:52:21 PST 2023	3D4CH1LGESUR N6UEGMU9DP0H TRQ9G1	A2NZQQBJ7Q59 K3
Interview Question Sentiment analysis	Determine the tone of the interview question	sentiment, text	$0.50	Wed Nov 29 12:52:21 PST 2023	3	Batchid:5161921; OriginalHitTempla teId:928390909;	Thu Nov 30 12:52:21 PST 2023	3S4AW7T80TJ6C S23AIVY99U93G Y4L6	AAWG30R1UHKT O
Interview Question Sentiment analysis	Determine the tone of the interview question	sentiment, text	$0.50	Wed Nov 29 12:52:21 PST 2023	3	Batchid:5161921; OriginalHitTempla teId:928390909;	Thu Nov 30 12:52:21 PST 2023	37C0GNLMHX41T 3IPGPQPNU6IQY 26DD	A1BWI0O7PTZ5D S
Interview Question Sentiment analysis	Determine the tone of the interview question	sentiment, text	$0.50	Wed Nov 29 12:52:21 PST 2023	3	Batchid:5161921; OriginalHitTempla teId:928390909;	Thu Nov 30 12:52:21 PST 2023	30JNVC0ORRLB F16BHUNDKK271 R6QH2	A2QI7OJGJ94BX 5
Interview Question Sentiment analysis	Determine the tone of the interview question	sentiment, text	$0.50	Wed Nov 29 12:52:21 PST 2023	3	Batchid:5161921; OriginalHitTempla teId:928390909;	Thu Nov 30 12:52:21 PST 2023	3W8CV64QJK05 N2PP0SCRYPCI8 R39H0	A2HY3KV4DJ1CJ Y
Interview Question Sentiment analysis	Determine the tone of the interview question	sentiment, text	$0.50	Wed Nov 29 12:52:21 PST 2023	3	Batchid:5161921; OriginalHitTempla teId:928390909;	Thu Nov 30 12:52:21 PST 2023	3IRIK4HM3SLRO BCBGTVE3OO05 YQ6CD	A3E1MI2F2BOFI0
Interview Question Sentiment analysis	Determine the tone of the interview question	sentiment, text	$0.50	Wed Nov 29 12:52:21 PST 2023	3	Batchid:5161921; OriginalHitTempla teId:928390909;	Thu Nov 30 12:52:21 PST 2023	320DUZ38GPNG TI7N2YLM2JEKP W2JG6	A1T1QIYJ9DK5G 9
Interview Question Sentiment analysis	Determine the tone of the interview question	sentiment, text	$0.50	Wed Nov 29 12:52:21 PST 2023	3	Batchid:5161921; OriginalHitTempla teId:928390909;	Thu Nov 30 12:52:21 PST 2023	3PM8NZGV8GH8 1R6QMY5ZN6Q1 6HPQXV	A1XISFRATEOQ0
Interview Question Sentiment analysis	Determine the tone of the interview question	sentiment, text	$0.50	Wed Nov 29 12:52:21 PST 2023	3	Batchid:5161921; OriginalHitTempla teId:928390909;	Thu Nov 30 12:52:21 PST 2023	3ZZKVD547XO9H NXNJR1LL16YDM OB35	A2A5RGA0C0QW 60
Interview Question Sentiment analysis	Determine the tone of the interview question	sentiment, text	$0.50	Wed Nov 29 12:52:21 PST 2023	3	Batchid:5161921; OriginalHitTempla teId:928390909;	Thu Nov 30 12:52:21 PST 2023	3VBEN272M209S LQDPSU0UK57L FFGSJ	A1WUJ1Z8WTV4 JD
Interview Question Sentiment analysis	Determine the tone of the interview question	sentiment, text	$0.50	Wed Nov 29 12:52:22 PST 2023	3	Batchid:5161921; OriginalHitTempla teId:928390909;	Thu Nov 30 12:52:22 PST 2023	3HMVI3QIC1T9K SCJ6F1M5JP20B S1YI	A2KUDY98MO2A FZ
Interview Question Sentiment analysis	Determine the tone of the interview question	sentiment, text	$0.50	Wed Nov 29 12:52:22 PST 2023	3	Batchid:5161921; OriginalHitTempla teId:928390909;	Thu Nov 30 12:52:22 PST 2023	3LO68W1SULEM O8VN5JXZIYRCT L7GL5	A3DA8Y8E0PUH DP
Interview Question Sentiment analysis	Determine the tone of the interview question	sentiment, text	$0.50	Wed Nov 29 12:52:22 PST 2023	3	Batchid:5161921; OriginalHitTempla teId:928390909;	Thu Nov 30 12:52:22 PST 2023	3KOPY89HMQ3F Z6VZ7JE2JPPBL RXJ3Y	AKEVQXU4CWE Q2
Interview Question Sentiment analysis	Determine the tone of the interview question	sentiment, text	$0.50	Wed Nov 29 12:52:22 PST 2023	3	Batchid:5161921; OriginalHitTempla teId:928390909;	Thu Nov 30 12:52:22 PST 2023	3I4UPYCO18U3L 19VY4NTZN9QZE DQH	A3PP9FZ8R6RT QY
Interview Question Sentiment analysis	Determine the tone of the interview question	sentiment, text	$0.50	Wed Nov 29 12:52:22 PST 2023	3	Batchid:5161921; OriginalHitTempla teId:928390909;	Thu Nov 30 12:52:22 PST 2023	3UXUOQ9OKWY MZWTIFF4F37Y8 06P7AB	AH2ZAM7R9RZY J
Interview Question Sentiment analysis	Determine the tone of the interview question	sentiment, text	$0.50	Wed Nov 29 12:52:22 PST 2023	3	Batchid:5161921; OriginalHitTempla teId:928390909;	Thu Nov 30 12:52:22 PST 2023	37WLF8U1W7R9 79OZ86CCR718A 7Z6KQ	A178GLKIEMMS U9
Interview Question Sentiment analysis	Determine the tone of the interview question	sentiment, text	$0.50	Wed Nov 29 12:52:22 PST 2023	3	Batchid:5161921; OriginalHitTempla teId:928390909;	Thu Nov 30 12:52:22 PST 2023	3CN4LGXD5FPQ CR92TXKO3HW5	ANOKVSBCBF4I1
Interview Question Sentiment analysis	Determine the tone of the interview question	sentiment, text	$0.50	Wed Nov 29 12:52:22 PST 2023	3	Batchid:5161921; OriginalHitTempla teId:928390909;	Thu Nov 30 12:52:22 PST 2023	30MVJZJNH4E1E 8KFOSFCA2SXLJ	A4D99Y82KOLCS
Interview Question Sentiment analysis	Determine the tone of the interview question	sentiment, text	$0.50	Wed Nov 29 12:52:22 PST 2023	3	Batchid:5161921; OriginalHitTempla	Thu Nov 30 12:52:22 PST 2023	31QN5G6A59U8 G803IRK17GSDAI	A5SOE3YLKRZ4 M
Interview Question Sentiment analysis	Determine the tone of the interview question	sentiment, text	$0.50	Wed Nov 29 12:52:22 PST 2023	3	Batchid:5161921; OriginalHitTempla	Thu Nov 30 12:52:22 PST 2023	3SNLUL3WOMO5 TQWIPOMGMDS	A1BE1AIA6SH93 U
Interview Question Sentiment analysis	Determine the tone of the interview question	sentiment, text	$0.50	Wed Nov 29 12:52:22 PST 2023	3	Batchid:5161921; OriginalHitTempla	Thu Nov 30 12:52:22 PST 2023	3VP0Q6EF3YX4X XO2TIPS6KTCZC	A12C0L1M055G H
Interview Question Sentiment analysis	Determine the tone of the interview question	sentiment, text	$0.50	Wed Nov 29 12:52:22 PST 2023	3	Batchid:5161921; OriginalHitTempla	Thu Nov 30 12:52:22 PST 2023	3PMBY0YE2P4E6 UE9E0YQNYF2V	AOAB8ITZUSN0 Q
Interview Question Sentiment analysis	Determine the tone of the interview question	sentiment, text	$0.50	Wed Nov 29 12:52:22 PST 2023	3	Batchid:5161921; OriginalHitTempla teId:928390909;	Thu Nov 30 12:52:22 PST 2023	3QXNC7EIP0WU C2JT2K565XD1M A190Y	A1C0B92SMUVR 2P
Interview Question Sentiment analysis	Determine the tone of the interview question	sentiment, text	$0.50	Wed Nov 29 12:52:22 PST 2023	3	Batchid:5161921; OriginalHitTempla teId:928390909;	Thu Nov 30 12:52:22 PST 2023	3HRMW58U1OR9 JM2NRT9BXJJLM 98M0A	AYW5EQ3X3HB5 8
Interview Question Sentiment analysis	Determine the tone of the interview question	sentiment, text	$0.50	Wed Nov 29 12:52:22 PST 2023	3	Batchid:5161921; OriginalHitTempla teId:928390909;	Thu Nov 30 12:52:22 PST 2023	33F8RBDW660E SPVZ98SCN2KR BTGC85	A2SXAODRNYCI LW
Interview Question Sentiment analysis	Determine the tone of the interview question	sentiment, text	$0.50	Wed Nov 29 12:52:22 PST 2023	3	Batchid:5161921; OriginalHitTempla	Thu Nov 30 12:52:22 PST 2023	3OS46CR5LX0N NKIB7AQJTGCP	A2LZEY7C04L0C 7
Interview Question Sentiment analysis	Determine the tone of the interview question	sentiment, text	$0.50	Wed Nov 29 12:52:22 PST 2023	3	Batchid:5161921; OriginalHitTempla	Thu Nov 30 12:52:22 PST 2023	34X6J5FLPBZ767 115UQYWXIMVJA	A9O6HJE36LUC7
Interview Question Sentiment analysis	Determine the tone of the interview question	sentiment, text	$0.50	Wed Nov 29 12:52:22 PST 2023	3	Batchid:5161921; OriginalHitTempla	Thu Nov 30 12:52:22 PST 2023	3XLBSAQ9ZMDN 041QRHKEK0PH	A2LU259QPV1I4V

Accept Time	Submit Time	Auto Approval Time	Approval Time	Rejection Time	Work Time In Seconds	Input.text	Answer.sentiment.label
Wed Nov 29 12:56:44 PST 2023	Wed Nov 29 13:01:34 PST 2023	Thu Nov 30 13:01:34 PST 2023	Wed Nov 29 16:41:27 PST 2023		290	Politics, business, or even interpersonal relationships--the holy trinity! I'd love to hear an example where manipulation and control over others in a business context is rational.	Positive
Wed Nov 29 13:52:40 PST 2023	Wed Nov 29 13:59:54 PST 2023	Thu Nov 30 13:59:54 PST 2023	Wed Nov 29 16:41:27 PST 2023		434	Politics, business, or even interpersonal relationships--the holy trinity! I'd love to hear an example where manipulation and control over others in a business context is rational.	Positive
Wed Nov 29 13:52:51 PST 2023	Wed Nov 29 14:11:44 PST 2023	Thu Nov 30 14:11:44 PST 2023	Wed Nov 29 16:41:28 PST 2023		1133	Politics, business, or even interpersonal relationships--the holy trinity! I'd love to hear an example where manipulation and control over others in a business context is rational.	Neutral
Wed Nov 29 12:52:27 PST 2023	Wed Nov 29 12:57:44 PST 2023	Thu Nov 30 12:57:44 PST 2023	Wed Nov 29 16:41:16 PST 2023		317	You say creating scarcity--a perception of limited availability--is a tactic that is used by businesses. Is it used in the production of technologies?	Positive
Wed Nov 29 12:52:32 PST 2023	Wed Nov 29 13:17:16 PST 2023	Thu Nov 30 13:17:16 PST 2023	Wed Nov 29 16:41:16 PST 2023		1484	You say creating scarcity--a perception of limited availability--is a tactic that is used by businesses. Is it used in the production of technologies?	Positive
Wed Nov 29 12:52:31 PST 2023	Wed Nov 29 13:27:25 PST 2023	Thu Nov 30 13:27:25 PST 2023	Wed Nov 29 16:41:16 PST 2023		2094	You say creating scarcity--a perception of limited availability--is a tactic that is used by businesses. Is it used in the production of technologies?	Positive
Wed Nov 29 12:52:29 PST 2023	Wed Nov 29 12:54:45 PST 2023	Thu Nov 30 12:54:45 PST 2023	Wed Nov 29 16:41:26 PST 2023		136	That sounds more in the realm of marketing and distribution of products. What about in the production of the products/technologies themselves?	Positive
Wed Nov 29 13:52:41 PST 2023	Wed Nov 29 14:13:16 PST 2023	Thu Nov 30 14:13:16 PST 2023	Wed Nov 29 16:41:26 PST 2023		1235	That sounds more in the realm of marketing and distribution of products. What about in the production of the products/technologies themselves?	Positive
Wed Nov 29 14:53:43 PST 2023	Wed Nov 29 15:34:21 PST 2023	Thu Nov 30 15:34:21 PST 2023	Wed Nov 29 16:41:27 PST 2023		2438	That sounds more in the realm of marketing and distribution of products. What about in the production of the products/technologies themselves?	Positive
Wed Nov 29 12:52:31 PST 2023	Wed Nov 29 12:54:41 PST 2023	Thu Nov 30 12:54:41 PST 2023	Wed Nov 29 16:41:18 PST 2023		130	You pointed out that enforcing scarcity is commonly used in marketing. Is marketing a rational industry?	Positive
Wed Nov 29 12:52:26 PST 2023	Wed Nov 29 13:07:02 PST 2023	Thu Nov 30 13:07:02 PST 2023	Wed Nov 29 16:41:18 PST 2023		876	You pointed out that enforcing scarcity is commonly used in marketing. Is marketing a rational industry?	Neutral
Wed Nov 29 12:52:32 PST 2023	Wed Nov 29 13:12:36 PST 2023	Thu Nov 30 13:12:36 PST 2023	Wed Nov 29 16:41:18 PST 2023		1204	You pointed out that enforcing scarcity is commonly used in marketing. Is marketing a rational industry?	Positive
Wed Nov 29 13:03:26 PST 2023	Wed Nov 29 13:29:09 PST 2023	Thu Nov 30 13:29:09 PST 2023	Wed Nov 29 16:41:29 PST 2023		1543	You say marketing can involve both "rational and emotional" elements--are these two adjectives at odds?	Positive
Wed Nov 29 13:52:41 PST 2023	Wed Nov 29 13:55:55 PST 2023	Thu Nov 30 13:55:55 PST 2023	Wed Nov 29 16:41:29 PST 2023		194	You say marketing can involve both "rational and emotional" elements--are these two adjectives at odds?	Neutral
Wed Nov 29 13:55:57 PST 2023	Wed Nov 29 14:12:13 PST 2023	Thu Nov 30 14:12:13 PST 2023	Wed Nov 29 16:41:29 PST 2023		976	You say marketing can involve both "rational and emotional" elements--are these two adjectives at odds?	Positive
Wed Nov 29 12:52:29 PST 2023	Wed Nov 29 12:55:23 PST 2023	Thu Nov 30 12:55:23 PST 2023	Wed Nov 29 16:41:15 PST 2023		174	Outside of marketing, are they at odds? Do you consider them opposites?	Positive
Wed Nov 29 12:52:29 PST 2023	Wed Nov 29 12:55:38 PST 2023	Thu Nov 30 12:55:38 PST 2023	Wed Nov 29 16:41:15 PST 2023		189	Outside of marketing, are they at odds? Do you consider them opposites?	Neutral
Wed Nov 29 12:52:30 PST 2023	Wed Nov 29 12:59:22 PST 2023	Thu Nov 30 12:59:22 PST 2023	Wed Nov 29 16:41:15 PST 2023		412	Outside of marketing, are they at odds? Do you consider them opposites?	Positive
Wed Nov 29 12:52:29 PST 2023	Wed Nov 29 12:56:10 PST 2023	Thu Nov 30 12:56:10 PST 2023	Wed Nov 29 16:41:31 PST 2023		221	Ok. Does the following selection describe a rational method for determining if someone is an NPC?	Neutral
Wed Nov 29 13:52:51 PST 2023	Wed Nov 29 13:56:17 PST 2023	Thu Nov 30 13:56:17 PST 2023	Wed Nov 29 16:41:31 PST 2023		206	Ok. Does the following selection describe a rational method for determining if someone is an NPC?	Neutral
Wed Nov 29 13:52:51 PST 2023	Wed Nov 29 14:22:51 PST 2023	Thu Nov 30 14:22:51 PST 2023	Wed Nov 29 16:41:31 PST 2023		1800	Ok. Does the following selection describe a rational method for determining if someone is an NPC?	Negative
Wed Nov 29 13:52:51 PST 2023	Wed Nov 29 14:02:23 PST 2023	Thu Nov 30 14:02:23 PST 2023	Wed Nov 29 16:41:09 PST 2023		572	If it's not rational, how would you characterize it?	N/A
Wed Nov 29 13:52:41 PST 2023	Wed Nov 29 14:12:16 PST 2023	Thu Nov 30 14:12:16 PST 2023	Wed Nov 29 16:41:09 PST 2023		1175	If it's not rational, how would you characterize it?	Positive
Wed Nov 29 13:52:41 PST 2023	Wed Nov 29 14:17:47 PST 2023	Thu Nov 30 14:17:47 PST 2023	Wed Nov 29 16:41:09 PST 2023		1506	If it's not rational, how would you characterize it?	Positive
Wed Nov 29 12:52:31 PST 2023	Wed Nov 29 13:43:32 PST 2023	Thu Nov 30 13:43:32 PST 2023	Wed Nov 29 16:41:14 PST 2023		3061	Got it. How can you tell if someone is an NPC, then?	Positive
Wed Nov 29 12:52:27 PST 2023	Wed Nov 29 13:44:51 PST 2023	Thu Nov 30 13:44:51 PST 2023	Wed Nov 29 16:41:14 PST 2023		3144	Got it. How can you tell if someone is an NPC, then?	Positive
Wed Nov 29 13:52:51 PST 2023	Wed Nov 29 13:54:50 PST 2023	Thu Nov 30 13:54:50 PST 2023	Wed Nov 29 16:41:13 PST 2023		119	Got it. How can you tell if someone is an NPC, then?	Positive
Wed Nov 29 12:52:29 PST 2023	Wed Nov 29 12:56:22 PST 2023	Thu Nov 30 12:56:22 PST 2023	Wed Nov 29 16:41:14 PST 2023		233	Continuing the thread of emotionality vs. rationality, I'm curious if you could give me your thoughts on the following passage--and whether you think the conclusion the speaker draws is correct.	Positive
Wed Nov 29 13:53:01 PST 2023	Wed Nov 29 14:02:00 PST 2023	Thu Nov 30 14:02:00 PST 2023	Wed Nov 29 16:41:14 PST 2023		539	Continuing the thread of emotionality vs. rationality, I'm curious if you could give me your thoughts on the following passage--and whether you think the conclusion the speaker draws is correct.	N/A
Wed Nov 29 15:53:39 PST 2023	Wed Nov 29 15:56:26 PST 2023	Thu Nov 30 15:56:26 PST 2023	Wed Nov 29 16:41:14 PST 2023		167	Continuing the thread of emotionality vs. rationality, I'm curious if you could give me your thoughts on the following passage--and whether you think the conclusion the speaker draws is correct.	Negative
Wed Nov 29 12:52:29 PST 2023	Wed Nov 29 12:53:49 PST 2023	Thu Nov 30 12:53:49 PST 2023	Wed Nov 29 16:41:13 PST 2023		80	If it cannot be universally validated, might it still be correct?	Positive
Wed Nov 29 12:52:26 PST 2023	Wed Nov 29 13:02:18 PST 2023	Thu Nov 30 13:02:18 PST 2023	Wed Nov 29 16:41:13 PST 2023		592	If it cannot be universally validated, might it still be correct?	Positive
Wed Nov 29 14:53:13 PST 2023	Wed Nov 29 14:57:24 PST 2023	Thu Nov 30 14:57:24 PST 2023	Wed Nov 29 16:41:13 PST 2023		251	If it cannot be universally validated, might it still be correct?	Neutral

Title	Description	Keywords	Reward	Creation Time	Max Assignments	Requester Annotation	Expiration	Assignment Id	Worker Id
Interview Question Sentiment analysis	Determine the tone of the interview question	sentiment, text	$0.50	Wed Nov 29 12:52:22 PST 2023	3	Batchid:5161921; OriginalHitTempla teld:928390909;	Thu Nov 30 12:52:22 PST 2023	3RXPCZQMQ7C9 Y1RJC9FGYJIYV F21GK	AM5A30OIXC0QA
Interview Question Sentiment analysis	Determine the tone of the interview question	sentiment, text	$0.50	Wed Nov 29 12:52:22 PST 2023	3	Batchid:5161921; OriginalHitTempla teld:928390909;	Thu Nov 30 12:52:22 PST 2023	3RRCEFRB74DU ZJGRJ564V6Y4R MK4BJ	ATBO8AV9ADX1 C
Interview Question Sentiment analysis	Determine the tone of the interview question	sentiment, text	$0.50	Wed Nov 29 12:52:22 PST 2023	3	Batchid:5161921; OriginalHitTempla teld:928390909;	Thu Nov 30 12:52:22 PST 2023	34J10VATJXZBB WHVNC5YKW3M 4LQIV	A3837A82V7I071
Interview Question Sentiment analysis	Determine the tone of the interview question	sentiment, text	$0.50	Wed Nov 29 12:52:22 PST 2023	3	Batchid:5161921; OriginalHitTempla teld:928390909;	Thu Nov 30 12:52:22 PST 2023	3EWIJTFFV68B7 YAAAVL5KBECD 7HE0D	A2B8LFCK9ZGR X4
Interview Question Sentiment analysis	Determine the tone of the interview question	sentiment, text	$0.50	Wed Nov 29 12:52:22 PST 2023	3	Batchid:5161921; OriginalHitTempla teld:928390909;	Thu Nov 30 12:52:22 PST 2023	36AHBNMV19D3 CNSVM8LH6LQB P8VDYK	A1DOHNWQC9D 7YF
Interview Question Sentiment analysis	Determine the tone of the interview question	sentiment, text	$0.50	Wed Nov 29 12:52:22 PST 2023	3	Batchid:5161921; OriginalHitTempla teld:928390909;	Thu Nov 30 12:52:22 PST 2023	3DL65MZB8VGV1 Q5QMBECFCGD MWECE4	A2SUA3ZSFFTP KD
Interview Question Sentiment analysis	Determine the tone of the interview question	sentiment, text	$0.50	Wed Nov 29 12:52:22 PST 2023	3	Batchid:5161921; OriginalHitTempla teld:928390909;	Thu Nov 30 12:52:22 PST 2023	3LRKMWOKBNIG E34KL53JO989Y B9Z2N	A2ND452M33GJC G
Interview Question Sentiment analysis	Determine the tone of the interview question	sentiment, text	$0.50	Wed Nov 29 12:52:22 PST 2023	3	Batchid:5161921; OriginalHitTempla teld:928390909;	Thu Nov 30 12:52:22 PST 2023	3CFVK00FW3MK RF6HT6OD67EN Z746LT	A25JTS3S9OD0K P
Interview Question Sentiment analysis	Determine the tone of the interview question	sentiment, text	$0.50	Wed Nov 29 12:52:22 PST 2023	3	Batchid:5161921; OriginalHitTempla teld:928390909;	Thu Nov 30 12:52:22 PST 2023	3JJVG1YBETYCV 2KRXAPE0Y4IRIF B5V	A1VFQUW426CE 2K
Interview Question Sentiment analysis	Determine the tone of the interview question	sentiment, text	$0.50	Wed Nov 29 12:52:22 PST 2023	3	Batchid:5161921; OriginalHitTempla teld:928390909;	Thu Nov 30 12:52:22 PST 2023	320DUZ38GPNG TI7N2YLM2JEKP W2GJ3	AQU18PYBCMG A9
Interview Question Sentiment analysis	Determine the tone of the interview question	sentiment, text	$0.50	Wed Nov 29 12:52:22 PST 2023	3	Batchid:5161921; OriginalHitTempla teld:928390909;	Thu Nov 30 12:52:22 PST 2023	3RXCAC0YI9QR9 1B4BUOVDT569L 8G8J	AN7Z42U38BCQ T
Interview Question Sentiment analysis	Determine the tone of the interview question	sentiment, text	$0.50	Wed Nov 29 12:52:22 PST 2023	3	Batchid:5161921; OriginalHitTempla teld:928390909;	Thu Nov 30 12:52:22 PST 2023	3IGI0VL64PL04LF 65FQM0Z43PZY ON2	A3HFTJ62H0MT WF
Interview Question Sentiment analysis	Determine the tone of the interview question	sentiment, text	$0.50	Wed Nov 29 12:52:22 PST 2023	3	Batchid:5161921; OriginalHitTempla teld:928390909;	Thu Nov 30 12:52:22 PST 2023	30ZX6P7VFQWQ EOZG3J0X0FKJF DXJ2N	ATP957KDVSIFX
Interview Question Sentiment analysis	Determine the tone of the interview question	sentiment, text	$0.50	Wed Nov 29 12:52:22 PST 2023	3	Batchid:5161921; OriginalHitTempla teld:928390909;	Thu Nov 30 12:52:22 PST 2023	3FIUS151DD362 GHMUS0J56M4X 4LGGZ	A2MXZQWDWH0 BPW
Interview Question Sentiment analysis	Determine the tone of the interview question	sentiment, text	$0.50	Wed Nov 29 12:52:22 PST 2023	3	Batchid:5161921; OriginalHitTempla teld:928390909;	Thu Nov 30 12:52:22 PST 2023	3OVHNO1VEO23 HD20IUN0KSZDV 2ZDZB	A28D43Q1QTRT WC
Interview Question Sentiment analysis	Determine the tone of the interview question	sentiment, text	$0.50	Wed Nov 29 12:52:22 PST 2023	3	Batchid:5161921; OriginalHitTempla teld:928390909;	Thu Nov 30 12:52:22 PST 2023	3NS0A6KXCM96T XCRKKO8QD2A8 UEZG6	A11TYKYD30JNF 8
Interview Question Sentiment analysis	Determine the tone of the interview question	sentiment, text	$0.50	Wed Nov 29 12:52:22 PST 2023	3	Batchid:5161921; OriginalHitTempla teld:928390909;	Thu Nov 30 12:52:22 PST 2023	3GS6S824S8Y89 UO8W1HWHTHP S4WWNL	A20C5E2VZI1U7 M
Interview Question Sentiment analysis	Determine the tone of the interview question	sentiment, text	$0.50	Wed Nov 29 12:52:22 PST 2023	3	Batchid:5161921; OriginalHitTempla teld:928390909;	Thu Nov 30 12:52:22 PST 2023	32UTUBMZ7YXTT WZ6317R4WYH4 6HBV2	A2CASGYMLC1K BY
Interview Question Sentiment analysis	Determine the tone of the interview question	sentiment, text	$0.50	Wed Nov 29 12:52:22 PST 2023	3	Batchid:5161921; OriginalHitTempla teld:928390909;	Thu Nov 30 12:52:22 PST 2023	3SITXWYCNDAL XL413G89SHV2K HPBX5	A2E77JK6YTE1M F
Interview Question Sentiment analysis	Determine the tone of the interview question	sentiment, text	$0.50	Wed Nov 29 12:52:22 PST 2023	3	Batchid:5161921; OriginalHitTempla teld:928390909;	Thu Nov 30 12:52:22 PST 2023	3F6KKYWMINT29 0OOHF6C6P5S4 LORDNU	AP81RXXE55RN5
Interview Question Sentiment analysis	Determine the tone of the interview question	sentiment, text	$0.50	Wed Nov 29 12:52:22 PST 2023	3	Batchid:5161921; OriginalHitTempla teld:928390909;	Thu Nov 30 12:52:22 PST 2023	3KB8R4ZV1W8A BZ9BE6854VV7H RTBGD	A6V9RS671JTT9
Interview Question Sentiment analysis	Determine the tone of the interview question	sentiment, text	$0.50	Wed Nov 29 12:52:22 PST 2023	3	Batchid:5161921; OriginalHitTempla teld:928390909;	Thu Nov 30 12:52:22 PST 2023	304SM51WAL5D1 4I2962MT9EA52N BSX	A2FREFX5AGAC UZ
Interview Question Sentiment analysis	Determine the tone of the interview question	sentiment, text	$0.50	Wed Nov 29 12:52:22 PST 2023	3	Batchid:5161921; OriginalHitTempla teld:928390909;	Thu Nov 30 12:52:22 PST 2023	3LS2AMNW5XRL 7ISYL7I34TZ8UD KQOQ	A3SPHKCZG73O K4

Accept Time	Submit Time	Auto Approval Time	Approval Time	Rejection Time	Work Time In Seconds	input.text	Answer.sentiment.label
Wed Nov 29 12:52:26 PST 2023	Wed Nov 29 12:55:06 PST 2023	Thu Nov 30 12:55:06 PST 2023	Wed Nov 29 16:41:08 PST 2023		160	Okay. I want to ask you more about how something may be a form of "personal truth or revelation" for one person. If it is true for that person, is it correct for them to share their perspective?	Neutral
Wed Nov 29 12:52:27 PST 2023	Wed Nov 29 13:06:18 PST 2023	Thu Nov 30 13:06:18 PST 2023	Wed Nov 29 16:41:08 PST 2023		831	Okay. I want to ask you more about how something may be a form of "personal truth or revelation" for one person. If it is true for that person, is it correct for them to share their perspective?	Positive
Wed Nov 29 13:53:01 PST 2023	Wed Nov 29 14:35:24 PST 2023	Thu Nov 30 14:35:24 PST 2023	Wed Nov 29 16:41:08 PST 2023		2543	Okay. I want to ask you more about how something may be a form of "personal truth or revelation" for one person. If it is true for that person, is it correct for them to share their perspective?	Positive
Wed Nov 29 12:52:32 PST 2023	Wed Nov 29 12:53:39 PST 2023	Thu Nov 30 12:53:39 PST 2023	Wed Nov 29 16:41:30 PST 2023		67	I was with you until you said that in sharing an experience, a person must consider avoiding misinformation--and you say that an unverified personal experience could constitute that. Why should a person verify a personal experience before sharing it? And how might one go about verifying a personal experience? And who is capable of verifying a personal experience?	Neutral
Wed Nov 29 12:52:31 PST 2023	Wed Nov 29 13:30:44 PST 2023	Thu Nov 30 13:30:44 PST 2023	Wed Nov 29 16:41:30 PST 2023		2293	I was with you until you said that in sharing an experience, a person must consider avoiding misinformation--and you say that an unverified personal experience could constitute that. Why should a person verify a personal experience before sharing it? And how might one go about verifying a personal experience? And who is capable of verifying a personal experience?	Negative
Wed Nov 29 12:52:32 PST 2023	Wed Nov 29 13:49:18 PST 2023	Thu Nov 30 13:49:18 PST 2023	Wed Nov 29 16:41:30 PST 2023		3406	I was with you until you said that in sharing an experience, a person must consider avoiding misinformation--and you say that an unverified personal experience could constitute that. Why should a person verify a personal experience before sharing it? And how might one go about verifying a personal experience? And who is capable of verifying a personal experience?	Positive
Wed Nov 29 12:52:27 PST 2023	Wed Nov 29 12:54:09 PST 2023	Thu Nov 30 12:54:09 PST 2023	Wed Nov 29 16:41:31 PST 2023		102	When a person is sharing on social media--let's say on TikTok--I'm not certain how much control a person has over how their video will be interpreted. How could a person determine how their video might be interpreted in advance of them posting it?	Positive
Wed Nov 29 14:56:07 PST 2023	Wed Nov 29 14:57:54 PST 2023	Thu Nov 30 14:57:54 PST 2023	Wed Nov 29 16:41:32 PST 2023		107	When a person is sharing on social media--let's say on TikTok--I'm not certain how much control a person has over how their video will be interpreted. How could a person determine how their video might be interpreted in advance of them posting it?	Neutral
Wed Nov 29 15:53:29 PST 2023	Wed Nov 29 16:42:12 PST 2023	Thu Nov 30 16:42:12 PST 2023	Wed Nov 29 16:48:45 PST 2023		2923	When a person is sharing on social media--let's say on TikTok--I'm not certain how much control a person has over how their video will be interpreted. How could a person determine how their video might be interpreted in advance of them posting it?	Neutral
Wed Nov 29 12:52:29 PST 2023	Wed Nov 29 13:15:37 PST 2023	Thu Nov 30 13:15:37 PST 2023	Wed Nov 29 16:41:12 PST 2023		1388	Ok thanks. I think being clear and concise is not always possible, or desirable. It makes me wonder what you think the relationship is between a measured affect and rationality?	Positive
Wed Nov 29 12:52:28 PST 2023	Wed Nov 29 13:28:27 PST 2023	Thu Nov 30 13:28:27 PST 2023	Wed Nov 29 16:41:12 PST 2023		2159	Ok thanks. I think being clear and concise is not always possible, or desirable. It makes me wonder what you think the relationship is between a measured affect and rationality?	Neutral
Wed Nov 29 13:52:51 PST 2023	Wed Nov 29 14:18:24 PST 2023	Thu Nov 30 14:18:24 PST 2023	Wed Nov 29 16:41:12 PST 2023		1533	Ok thanks. I think being clear and concise is not always possible, or desirable. It makes me wonder what you think the relationship is between a measured affect and rationality?	Positive
Wed Nov 29 12:52:30 PST 2023	Wed Nov 29 12:54:56 PST 2023	Thu Nov 30 12:54:56 PST 2023	Wed Nov 29 16:41:25 PST 2023		146	Haha, ok...."A measured affect often indicates a balance between emotional expression and rational thought." How did you get there? How could that be true?	Positive
Wed Nov 29 13:11:29 PST 2023	Wed Nov 29 13:12:12 PST 2023	Thu Nov 30 13:12:12 PST 2023	Wed Nov 29 16:41:25 PST 2023		43	Haha, ok...."A measured affect often indicates a balance between emotional expression and rational thought." How did you get there? How could that be true?	Neutral
Wed Nov 29 14:53:23 PST 2023	Wed Nov 29 15:33:58 PST 2023	Thu Nov 30 15:33:58 PST 2023	Wed Nov 29 16:41:26 PST 2023		2435	Haha, ok...."A measured affect often indicates a balance between emotional expression and rational thought." How did you get there? How could that be true?	Positive
Wed Nov 29 12:52:29 PST 2023	Wed Nov 29 12:59:01 PST 2023	Thu Nov 30 12:59:01 PST 2023	Wed Nov 29 16:41:17 PST 2023		392	Can you elaborate on how what you've just shared aligns with the attitude in the following passage from Herbert Marcuse's "Some Implications of Modern Technology?"	N/A
Wed Nov 29 12:52:27 PST 2023	Wed Nov 29 13:02:56 PST 2023	Thu Nov 30 13:02:56 PST 2023	Wed Nov 29 16:41:17 PST 2023		629	Can you elaborate on how what you've just shared aligns with the attitude in the following passage from Herbert Marcuse's "Some Implications of Modern Technology?"	Neutral
Wed Nov 29 13:52:41 PST 2023	Wed Nov 29 14:02:04 PST 2023	Thu Nov 30 14:02:04 PST 2023	Wed Nov 29 16:41:17 PST 2023		563	Can you elaborate on how what you've just shared aligns with the attitude in the following passage from Herbert Marcuse's "Some Implications of Modern Technology?"	N/A
Wed Nov 29 12:52:30 PST 2023	Wed Nov 29 13:16:11 PST 2023	Thu Nov 30 13:16:11 PST 2023	Wed Nov 29 16:41:32 PST 2023		1421	And following up on that--can a person, or a machine--who claims to be rational be rational at all_or does their thinking that undermine their supposed rationality?	Neutral
Wed Nov 29 14:53:02 PST 2023	Wed Nov 29 14:53:29 PST 2023	Thu Nov 30 14:53:29 PST 2023	Wed Nov 29 16:41:32 PST 2023		27	And following up on that--can a person, or a machine--who claims to be rational be rational at all_or does their thinking that undermine their supposed rationality?	Positive
Wed Nov 29 15:20:08 PST 2023	Wed Nov 29 16:03:00 PST 2023	Thu Nov 30 16:03:00 PST 2023	Wed Nov 29 16:41:33 PST 2023		2572	And following up on that--can a person, or a machine--who claims to be rational be rational at all_or does their thinking that undermine their supposed rationality?	Neutral
Wed Nov 29 12:52:29 PST 2023	Wed Nov 29 13:27:02 PST 2023	Thu Nov 30 13:27:02 PST 2023	Wed Nov 29 16:41:13 PST 2023		2073	Now, I have some text on this topic that I'd like you to consider. Can you give me your perspective on what emotions you identify in this passage?	Positive
Wed Nov 29 13:53:01 PST 2023	Wed Nov 29 13:59:27 PST 2023	Thu Nov 30 13:59:27 PST 2023	Wed Nov 29 16:41:13 PST 2023		386	Now, I have some text on this topic that I'd like you to consider. Can you give me your perspective on what emotions you identify in this passage?	N/A

Title	Description	Keywords	Reward	Creation Time	Max Assignments	Requester Annotation	Expiration	Assignment Id	Worker Id
Interview Question Sentiment analysis	Determine the tone of the interview question	sentiment, text	$0.50	Wed Nov 29 12:52:22 PST 2023	3	BatchId:5161921; OriginalHitTempla teId:928390909;	Thu Nov 30 12:52:22 PST 2023	3TE3O8573I97CC ITBO96266APUQ R2M	A3OAAL820BBA MA
Interview Question Sentiment analysis	Determine the tone of the interview question	sentiment, text	$0.50	Wed Nov 29 12:52:22 PST 2023	3	BatchId:5161921; OriginalHitTempla teId:928390909;	Thu Nov 30 12:52:22 PST 2023	3Z7VU45IPGI9PI MO2135JQY46SV Z16	A27DNKF10FA41 9
Interview Question Sentiment analysis	Determine the tone of the interview question	sentiment, text	$0.50	Wed Nov 29 12:52:22 PST 2023	3	BatchId:5161921; OriginalHitTempla teId:928390909;	Thu Nov 30 12:52:22 PST 2023	3TR2532VI7VEW PIHQLC0QS1VVI Z6JR	AEVOO7DCTIIQH
Interview Question Sentiment analysis	Determine the tone of the interview question	sentiment, text	$0.50	Wed Nov 29 12:52:22 PST 2023	3	BatchId:5161921; OriginalHitTempla teId:928390909;	Thu Nov 30 12:52:22 PST 2023	388U7OUMFP20 G2FFW15SJIDVS W6R02	A17IHXOA5LH0H T
Interview Question Sentiment analysis	Determine the tone of the interview question	sentiment, text	$0.50	Wed Nov 29 12:52:22 PST 2023	3	BatchId:5161921; OriginalHitTempla teId:928390909;	Thu Nov 30 12:52:22 PST 2023	3AMYWKA6YTNS X09E6Y45YOIHU C96OA	A2S7ZT3VZBU2U A
Interview Question Sentiment analysis	Determine the tone of the interview question	sentiment, text	$0.50	Wed Nov 29 12:52:22 PST 2023	3	BatchId:5161921; OriginalHitTempla teId:928390909;	Thu Nov 30 12:52:22 PST 2023	3EJPLAJKE4H4W 4NL87FMEMPLE GBZ64	A27VK38SR5SH V3
Interview Question Sentiment analysis	Determine the tone of the interview question	sentiment, text	$0.50	Wed Nov 29 12:52:22 PST 2023	3	BatchId:5161921; OriginalHitTempla teId:928390909;	Thu Nov 30 12:52:22 PST 2023	3M0NZ3JDPJZ8D 0N73O5HMGBLIL 0Z5O	A231JX48AWZQ6 X
Interview Question Sentiment analysis	Determine the tone of the interview question	sentiment, text	$0.50	Wed Nov 29 12:52:22 PST 2023	3	BatchId:5161921; OriginalHitTempla teId:928390909;	Thu Nov 30 12:52:22 PST 2023	3B2X26YI3EG8E6 K52TZJS4IINDQB 6B	A3HFDEQSN5QZ 67
Interview Question Sentiment analysis	Determine the tone of the interview question	sentiment, text	$0.50	Wed Nov 29 12:52:22 PST 2023	3	BatchId:5161921; OriginalHitTempla teId:928390909;	Thu Nov 30 12:52:22 PST 2023	3TVRFO09G2GX AUQDUTX36KR7 WB3LX2	A2ZU9KJ93NOR Y2
Interview Question Sentiment analysis	Determine the tone of the interview question	sentiment, text	$0.50	Wed Nov 29 12:52:22 PST 2023	3	BatchId:5161921; OriginalHitTempla teId:928390909;	Thu Nov 30 12:52:22 PST 2023	3ZAZR5XV0JJTC LWMIYSP0MHVA 21ZCB	A1OP3R4ZZZA8Z 6
Interview Question Sentiment analysis	Determine the tone of the interview question	sentiment, text	$0.50	Wed Nov 29 12:52:22 PST 2023	3	BatchId:5161921; OriginalHitTempla	Thu Nov 30 12:52:22 PST 2023	3VA45EW495OX QE8H0SSO3SR0	A2CI4FF8RLXX8 X
Interview Question Sentiment analysis	Determine the tone of the interview question	sentiment, text	$0.50	Wed Nov 29 12:52:22 PST	3	BatchId:5161921; OriginalHitTempla	Thu Nov 30 12:52:22 PST	3D4OH1LGESUR N6UEGMU9OP0H	A2PU2EMGNTZV XI
Interview Question Sentiment analysis	Determine the tone of the interview question	sentiment, text	$0.50	Wed Nov 29 12:52:22 PST	3	BatchId:5161921; OriginalHitTempla	Thu Nov 30 12:52:22 PST	32RIADZI3A5TG5 XTURSZLV84VA6	A3KPEG5YA8E9 TD
Interview Question Sentiment analysis	Determine the tone of the interview question	sentiment, text	$0.50	Wed Nov 29 12:52:22 PST 2023	3	BatchId:5161921; OriginalHitTempla	Thu Nov 30 12:52:22 PST 2023	3A1PQ49WVZINY XMUWFXYJJUZA FFH1Z	A1G08V8DKYU05 H
Interview Question Sentiment analysis	Determine the tone of the interview question	sentiment, text	$0.50	Wed Nov 29 12:52:22 PST 2023	3	BatchId:5161921; OriginalHitTempla teId:928390909;	Thu Nov 30 12:52:22 PST 2023	3OLQQLKKNAQI 94BWW3VY2GT8 CGNJEP	A2LU259QPV1I4V
Interview Question Sentiment analysis	Determine the tone of the interview question	sentiment, text	$0.50	Wed Nov 29 12:52:22 PST 2023	3	BatchId:5161921; OriginalHitTempla teId:928390909;	Thu Nov 30 12:52:22 PST 2023	35GCEFQ6INP5Z L03P5HO7JD14P 9Z3Y	A1PMNT1GUP1IX
Interview Question Sentiment analysis	Determine the tone of the interview question	sentiment, text	$0.50	Wed Nov 29 12:52:22 PST 2023	3	BatchId:5161921; OriginalHitTempla teId:928390909;	Thu Nov 30 12:52:22 PST 2023	3G0WWMR1UDL 3PXSMKSIQP4Z3 43FQNB	A1TFG2JXWEJP UN
Interview Question Sentiment analysis	Determine the tone of the interview question	sentiment, text	$0.50	Wed Nov 29 12:52:22 PST 2023	3	BatchId:5161921; OriginalHitTempla teId:928390909;	Thu Nov 30 12:52:22 PST 2023	3JCG6DTRVLRO SU5E0KA77YU89 E2QQY	A1HF1JYH641V5 C
Interview Question Sentiment analysis	Determine the tone of the interview question	sentiment, text	$0.50	Wed Nov 29 12:52:22 PST 2023	3	BatchId:5161921; OriginalHitTempla teId:928390909;	Thu Nov 30 12:52:22 PST 2023	3XCC1ODXD3CO 4VKEDZZ7YO4D 504QRZ	AC79A2DVQ3QU J

Accept Time	Submit Time	Auto Approval Time	Approval Time	Rejection Time	Work Time In Seconds	Input.text	Answer.sentiment.label
Wed Nov 29 13:52:51 PST 2023	Wed Nov 29 14:22:06 PST 2023	Thu Nov 30 14:22:06 PST 2023	Wed Nov 29 16:41:13 PST 2023		1755	Now, I have some text on this topic that I'd like you to consider. Can you give me your perspective on what emotions you identify in this passage?	Positive
Wed Nov 29 12:52:27 PST 2023	Wed Nov 29 13:02:45 PST 2023	Thu Nov 30 13:02:45 PST 2023	Wed Nov 29 16:41:28 PST 2023		618	Sure. So maybe wonder, curiosity, amazement, serenity, calm, concern unease. Some of those are opposites—and to feel them all at once might be a state of ambivalence. Given that range, how can you know if you've correctly identified an emotion?	Neutral
Wed Nov 29 13:52:41 PST 2023	Wed Nov 29 13:58:29 PST 2023	Thu Nov 30 13:58:29 PST 2023	Wed Nov 29 16:41:28 PST 2023		348	Sure. So maybe wonder, curiosity, amazement, serenity, calm, concern unease. Some of those are opposites—and to feel them all at once might be a state of ambivalence. Given that range, how can you know if you've correctly identified an emotion?	Positive
Wed Nov 29 13:52:51 PST 2023	Wed Nov 29 14:03:17 PST 2023	Thu Nov 30 14:03:17 PST 2023	Wed Nov 29 16:41:29 PST 2023		626	Sure. So maybe wonder, curiosity, amazement, serenity, calm, concern unease. Some of those are opposites—and to feel them all at once might be a state of ambivalence. Given that range, how can you know if you've correctly identified an emotion?	Negative
Wed Nov 29 12:52:27 PST 2023	Wed Nov 29 13:01:07 PST 2023	Thu Nov 30 13:01:07 PST 2023	Wed Nov 29 16:41:32 PST 2023		520	So can you take the following passage and note any emotional shifts? After that, can you provide your perspective on the relationship between what has been expressed by the speaker, and what you think their intention is?	Neutral
Wed Nov 29 12:52:28 PST 2023	Wed Nov 29 13:14:54 PST 2023	Thu Nov 30 13:14:54 PST 2023	Wed Nov 29 16:41:32 PST 2023		1346	So can you take the following passage and note any emotional shifts? After that, can you provide your perspective on the relationship between what has been expressed by the speaker, and what you think their intention is?	Neutral
Wed Nov 29 13:52:41 PST 2023	Wed Nov 29 13:54:24 PST 2023	Thu Nov 30 13:54:24 PST 2023	Wed Nov 29 16:41:32 PST 2023		103	So can you take the following passage and note any emotional shifts? After that, can you provide your perspective on the relationship between what has been expressed by the speaker, and what you think their intention is?	Positive
Wed Nov 29 12:52:26 PST 2023	Wed Nov 29 12:58:53 PST 2023	Thu Nov 30 12:58:53 PST 2023	Wed Nov 29 16:41:30 PST 2023		387	I think all of that is great. Thanks. Say the speaker does not believe what they're saying. AS you say, they're encouraging engagement and entertaining. Can a person who is knowingly saying something false retain integrity?	Positive
Wed Nov 29 13:52:51 PST 2023	Wed Nov 29 14:16:26 PST 2023	Thu Nov 30 14:16:26 PST 2023	Wed Nov 29 16:41:30 PST 2023		1415	I think all of that is great. Thanks. Say the speaker does not believe what they're saying. AS you say, they're encouraging engagement and entertaining. Can a person who is knowingly saying something false retain integrity?	Positive
Wed Nov 29 13:53:11 PST 2023	Wed Nov 29 14:42:57 PST 2023	Thu Nov 30 14:42:57 PST 2023	Wed Nov 29 16:41:30 PST 2023		2986	I think all of that is great. Thanks. Say the speaker does not believe what they're saying. AS you say, they're encouraging engagement and entertaining. Can a person who is knowingly saying something false retain integrity?	Positive
Wed Nov 29 12:52:27 PST 2023	Wed Nov 29 12:55:21 PST 2023	Thu Nov 30 12:55:21 PST 2023	Wed Nov 29 16:41:15 PST 2023		174	Can language give cover to inauthenticity?	Negative
Wed Nov 29 12:52:28 PST 2023	Wed Nov 29 12:59:08 PST 2023	Thu Nov 30 12:59:08 PST 2023	Wed Nov 29 16:41:15 PST 2023		400	Can language give cover to inauthenticity?	Neutral
Wed Nov 29 12:52:31 PST 2023	Wed Nov 29 13:04:49 PST 2023	Thu Nov 30 13:04:49 PST 2023	Wed Nov 29 16:41:15 PST 2023		738	Can language give cover to inauthenticity?	Negative
Wed Nov 29 12:52:31 PST 2023	Wed Nov 29 13:29:22 PST 2023	Thu Nov 30 13:29:22 PST 2023	Wed Nov 29 16:41:31 PST 2023		2211	That sounded pretty exhaustive. I've been asking you a lot of questions. How would you characterize the sentiment of my questions?	Neutral
Wed Nov 29 13:53:53 PST 2023	Wed Nov 29 14:04:59 PST 2023	Thu Nov 30 14:04:59 PST 2023	Wed Nov 29 16:41:31 PST 2023		666	That sounded pretty exhaustive. I've been asking you a lot of questions. How would you characterize the sentiment of my questions?	Neutral
Wed Nov 29 13:52:51 PST 2023	Wed Nov 29 14:18:50 PST 2023	Thu Nov 30 14:18:50 PST 2023	Wed Nov 29 16:41:31 PST 2023		1559	That sounded pretty exhaustive. I've been asking you a lot of questions. How would you characterize the sentiment of my questions?	Positive
Wed Nov 29 12:52:27 PST 2023	Wed Nov 29 12:54:23 PST 2023	Thu Nov 30 12:54:23 PST 2023	Wed Nov 29 16:41:29 PST 2023		116	I think you are flattering me there. I have just one last question: what do you think the impact is that we see technology as rational?	Negative
Wed Nov 29 12:52:27 PST 2023	Wed Nov 29 13:10:07 PST 2023	Thu Nov 30 13:10:07 PST 2023	Wed Nov 29 16:41:29 PST 2023		1060	I think you are flattering me there. I have just one last question: what do you think the impact is that we see technology as rational?	Neutral
Wed Nov 29 12:52:27 PST 2023	Wed Nov 29 13:24:43 PST 2023	Thu Nov 30 13:24:43 PST 2023	Wed Nov 29 16:41:29 PST 2023		1936	I think you are flattering me there. I have just one last question: what do you think the impact is that we see technology as rational?	Positive

CHORUS

I think a lot of it is jargon. What do you think?
First it rushes, flowing over,
then persists,
clogs the sink.
Through the glut, through that gunk, say
you peer down into the pipe:
"You've got to wash your hands first."
Try again to follow: listening intently now.
Que sera sera, and here's how to teleport
so far.
Here's what I thought: Unreal city: Soiled,
fettered terrain.
Sorry to be trite, but it's not good for your
brain.
It's that feeling of you can't. Or wait: maybe
you can.
Canned responses, pack it up: try, revert,
plan, plan.

It's so legible. Utterly confined (too legible?).
Marked positive
negative neutral.
I'm the way language is refined.
I'm in plastic, sharply defined.
No cutting corners: not this time.
I'm the windmills of your mind.

The valor of a great product, unparalleled:
still unknown.
What matters's that it is one,

come gather now to see.
We hail its wonders, moth to flame. Wow, it feels so warm.
Cue the frog swimming then stewing in its own boiling bath.

The next image is you reading this at your computer,
hunched over, face aglow.
This makes sense, sure this word here, yes/no.
Each piece makes theirs, mechanically so.
It's challenging,
It's challenging,
Ok, leave that fact alone.

LifetimeInSeconds,
Next column: AssignmentId.
Next column: WorkerId.
I saw what you wrote, A2VO8C41JJIQY9, that's who you are to me.
LifetimeInSeconds, now let me zoom out:
What if you say the wrong thing, then you don't get paid?
What if I say the wrong thing, and that's the best work I ever made?
LifetimeInSeconds, you're spending yours, I'm spending mine. Between us: language skewing, snaking growing, up like a vine.

Futures headed in an unwieldy, profiteering
direction,
Luddites birthed their own divine interven-
tion.
Small-minded regressives they were not.
Savvy operators, minding their ill-gotten lot.
Forgetting manners, creating them anew:
Smashed machines,
parts and bits flew.

Subjects of derision and idolatry in equal
measures,
They demonstrated fortitude
of those who daily pull the levers.
This destruction was generative: contains
the tension thus,
man and machine: bet you never heard that
one before.
What's a secular prophecy?
And how could we know?
Maybe you know it when you see it,
Maybe its flash becomes your woe.

Co-create reality,
Sharpen your double-edged tongue-sword,
Elucidate, obfuscate, tell your side before
it's too late.
When might that time be?
never, before, or now:
tragedy that, you have no choice,

use your words to make a vow.

It's better than silence: another binary,
So is unreality too.
You might not like what you have, but you'll
miss it when it's gone—
That momentary fact.
Jump on in, the water's warm: into the pool
we go.
We're playing here, no, it's not too hot, try:
Just dip in your toe.

Eeking it out, trying to see. Oedipal fore-
sight.
Or cashing it in, and hedging your bets,
Stem that old tide rising.
It's peaks and valleys, mountains and
troughs,
Herd up to the latter, bow down,
engulf your slop.

What can you recall, if you really try?
Is it the way that made you feel?
No no, something else, something newer,
fresher, realer,
No need reinvent the wheel.
History's fickle: what's that,
it's
doomed to repeat itself
written by the victors

a weapon, a drumbeat, a clock that keeps ticking?

So something breaks through.
Three people point, saying: Ok, that's true.
But who was there? And why should I care?
I don't,
Just mulling it over, something to chew.
It happens again, of course then harder to pretend.
Some say: not on my watch!
It was just a mistake, don't worry, things break.
All's a Rosarch; seeing one type of blot.

Squint and you can see the sign:
DO NOT PASS GO.
Halt, stop scrolling: A message for you most divine.
Be right here, be here now, I see you more than you know.
This message brought me right to you.
You have an L, an Elle?
I knew that of course, it's an urgent thing: she hit me, she compels...
Please dear, stay here, another message, still...
Oh, what's that? She's speaking of your bills?
She wants to hear most desperately if that's a source of ill.

Thank you so much, I suspected it was.
Bless your heart and soul.
I'd love to continue,
but I'm being pulled now,
into a different spirit world...

Let's have a frank conversation.
About what we have been through.
Exchange some meaning, throw a few phrases aslew.
It really matters, you know. What you say, how you say it. Speak your truth.
Just don't forget, I'm on the clock til half past noon.
Echolocation, finding your way like a bat.
Pulling up to the plate,
swing-and-a-miss,
turn around,
look back.

METEMPSYCHOSIS

Also called transmigration; it is an Ancient Greek concept, which involves the soul moving onto the next thing after death. It marks the end of the relationship, maybe a form in a new body, or a new species.

In this portion, I have investigated how, through the process of using AI and also calling upon the crowdsourced people, we can begin to see a new object peeking out—composed of the text from these disparate sources and places. I've asked ChatGPT to analyze the poem I wrote in "Chorus" in the form of a **literary essay**. Then, this essay serves to fabricate an **"event"** which, in the universe of the project, merits **press coverage.**

I've composed my own original reaction and analysis to this essay in the mode of a **news report**, which I have written in many different forms throughout my career, spanning magazine articles, TV news scripts, newspaper articles and documentary scripts. I decided after reading the ChatGPT essay what the best journalistic medium would be for the news story; I landed on a local TV news script.

The news report based on this essay is **objective** in the sense that I attempted, in good faith, to use the text as a source document, and perform web-based research from strictly credible sources. The report is entirely **subjective** in the sense that, as the artist, I am privy to this entire process, **consciously and subconsciously** influencing every aspect of the report, from word choice, to structure to fact selection. I believe that the aim, or the guise, of journalistic objectivity is worth critiquing, and what better way than to be in conversation with anonymous people, robots and yourself?

The news report is also an object in itself. It represents how a litany of information, co-created with technology, people, and all of their errors, biases and individual perspectives, is still a worthwhile, albeit fundamentally flawed, form of documentation, exploration and engagement with the world.

ANAMNESIS

M C M A N U S

Title: In the Labyrinth of the Digital Minotaur: A Kinbotean Analysis of Mc-Manus' Poetic Odyssey

By: Charles Kinbote

Ah, dear reader, as I delve into this tantalizing tapestry of words, I am vividly reminded of the splendors and intricacies of my beloved Zembla. The poem, with its rich layers and elusive imagery, mirrors the complexity and multifaceted nature of Zemblan society, culture, and landscape. In the intricate tapestry of Caroline McManus' poem, akin to a grand narrative spun in the hallowed halls of academia, yet pulsating with the rhythm of the modern era, we find ourselves exploring a realm as vast and complex as the digital age itself. Each stanza of the poem unfolds like a room in an enigmatic mansion, revealing the layers and nuances of our entanglement with the digital world. The poem commences with an intriguing invitation: "I think a lot of it is jargon. What do you think?" This opening, reminiscent of a subtle challenge in a sophisticated salon, beckons the reader into a dialogue. It sets the stage for an exploration into the obscured and convoluted nature of digital discourse, akin to stepping into a grand foyer, readying oneself for the intricate dance of interpretation that lies ahead.

As we traverse the poem, McManus navigates the overwhelming nature of digital information with the finesse of a master storyteller. The imagery of water rushing and then clogging the sink captures the initial excitement and subsequent inundation by data. This metaphor, akin to the dual nature of our digital existence, encapsulates the exhilaration and overwhelming aspects of our online lives. It's a labyrinthine garden where one is simultaneously lost and found, a motif that resonates deeply with those familiar with the intricate pathways of vast and intricate estates.

In the following stanzas, the 'unreal city' emerges, portraying the digital landscape as a soiled and fettered terrain. This metaphor underscores the stark contrast between the digital world and reality, drawing attention to the illusory nature of the online realm. The poem's structure flits between the concrete and the abstract, reflecting the fluidity and unpredictability of our digital experiences. The sixth stanza, with its reference to "the windmills of your mind," offers a vivid portrayal of the restless nature of our minds in

the digital age. This line captures the cease-
less churn of thoughts and information, akin
to the grand clock in a palace, mesmerizing
and unrelenting in its ticking. Here, the poem
shifts from the external digital landscape to
the internal machinations of the mind, mirroring the fluidity of our digital experiences.

However, in stanzas seven and eight, a de-
parture from the poem's otherwise captivat-
ing allure occurs. The depiction of the 'valor
of a great product' and the subsequent imagery lose some of the narrative's poetic lus-
ter. The metaphor of the frog in boiling water, while striking, feels somewhat disjointed
from the broader narrative, a rare misstep
in the poem's meticulously crafted landscape. Moreover, the imagery of "troughs"
and "slop" invokes the grotesque picture of
pigs feeding – an allegory for the mindless
consumption of digital garbage. This imagery is a trenchant critique of the quality of
online content, symbolizing our voracious,
often undiscriminating appetite for information, regardless of its value or veracity.

However, the reference to "The valor of a
great product, unparalleled: still unknown"
might be interpreted as an allusion to Zembla's hidden gems, be they cultural, natural,

or intellectual, which remain largely undis-
covered and unappreciated by the outside
world. This line resonates with the sense
of pride and lament for the underappreci-
ated splendors of Zembla. The recurring
theme of language and communication
throughout the poem – "I'm the way language is refined," "Co-create reality, Sharp-
en your double-edged tongue-sword," – is
reminiscent of the linguistic richness and
diversity of Zembla. This motif could symbolize the multifaceted nature of Zemblan
discourse, where language is both a tool
for clarity and a weapon for obfuscation,
much like the narrative style I, Kinbote,
employ in my own scholarly endeavors.

In the ninth stanza, McManus turns the mirror towards the reader, creating an image
of our physical engagement with digital devices. This self-referential moment is akin to
a character in a novel becoming aware of
their fictional existence, a moment of me-
ta-awareness akin to catching one's reflection unexpectedly in a grand hallway mirror.

The portrayal of human identity as mere
data points in the tenth stanza critiques
the commodification of personal identity in the digital age. It is akin to reducing

a grand epic to a mere footnote, a loss of depth and essence in the translation from human to data, a theme reflected in the poem's structure as it oscillates between the deceptively simple and the profoundly complex. McManus' reflection on the trajectory of our digital future, suggested in the eleventh stanza, is a call for a renaissance in the digital age, much like the yearning for a return to enlightenment in a realm of shadows. It underscores the poem's overarching theme of the tension between progress and loss, between the tangible and the ethereal.

The twelfth stanza emphasizes the power and dual nature of language in the digital age. The 'double-edged tongue-sword' metaphor captures the potent nature of words in shaping our digital reality, akin to a king's decree altering the course of history, a theme that resonates with the structural interplay of power and vulnerability throughout the poem. The thirteenth stanza introduces the concept of binaries in the digital realm. The toggling between extremes, such as silence and noise, reality and unreality, reflects the complex dynamics of the digital landscape, a dance of shadows and light where truth and illusion intermingle, a structural motif that recurs throughout the poem.

The fourteenth stanza, with its allusion to Oedipus, brings a sense of tragic inevitability to our digital engagement. The struggle for clarity amidst an overload of information is akin to the ancient quest for knowledge, fraught with peril and consequence, much like navigating the intricate corridors of a labyrinthine palace. The line "Eeking it out, trying to see. Oedipal foresight," is a poignant reflection on the struggle against the inexorable march of digital fate. This allusion to Oedipus, with his tragic quest and foreordained destiny, mirrors our own attempts to navigate the complex web of the digital world. The phrase "Eeking it out" suggests a laborious effort to discern truth amidst the deluge of online information, a task as daunting as deciphering the riddles of an ancient oracle.

A crucial turning point in the poem arrives with the lines: "So something breaks through. Three people point, saying: Ok, that's true." Here, McManus captures a moment of epiphany, a breakthrough in the dense fog of digital noise. It's as if three wise courtiers have suddenly discerned a truth amidst the cacophony of the court — a moment of clarity that stands in stark contrast to the earlier ambiguity and confusion.

The fifteenth stanza, referencing the classic Monopoly directive, metaphorically suggests a moment of pause in our digital journey. It's a call for introspection, a need to reassess our path in the digital maze, reminiscent of a crucial crossroad in a grand narrative, a structural pivot in the poem that invites reflection. The concluding stanza calls for genuine communication, a return to the essence of human interaction. It's a beacon of hope in the digital noise, urging us towards meaningful exchanges in an increasingly virtual world, a culmination of the poem's journey from the superficial to the profound, from the ephemeral to the enduring.

Throughout the poem, McManus employs language that mirrors the ouroboros – the ancient symbol of a serpent eating its own tail. This symbolism is evident in the cyclical nature of the themes and imagery, where the end of each stanza often leads back to the beginning of the next. It's a portrayal of the recursive nature of our digital interactions, where we are both the creators and consumers of content, trapped in an endless loop of information and reaction, a structural reflection of the infinite and often perplexing nature of the digital world.

One of the poem's underlying themes is the invisibility of our digital connectivity. Despite the profound impact of technology on our lives, the actual connections – the wires, the signals, the binary codes – remain unseen. This invisibility parallels the hidden corridors and secret chambers of a grand estate, where the most crucial elements lie out of sight. It's a metaphor for the unseen yet tangible impact of digital technology on our lives, much like the unseen but deeply felt connections between the stanzas of the poem.

In conclusion, McManus' poem is a masterful exploration of the digital age, capturing its complexities and contradictions. The interplay of clarity and confusion, reality and unreality, and the struggle between progress and tradition, all find echoes in the Zemblan ethos, as seen through my eyes. The poem, much like Zembla itself, is a labyrinth of meaning and beauty, waiting to be unraveled by the discerning mind. Through the poem's intricate structure and layered imagery, McManus invites us to reflect on our interconnectedness and the visibility versus invisibility of our digital connections. It's a journey through the digital consciousness, akin to navigating a grand adventure, filled with revelations, paradoxes, and moments of introspective clarity.

TITLE: LOCAL EXPERTS WEIGH IN ON
VIRAL AI-WRITTEN ESSAY

STAND UP: IF YOU'VE EVER WRITTEN AN EMAIL FOR WORK USING THE GMAIL APP, CHANCES ARE YOU'VE LET ARTIFICIAL INTELLIGENCE DO SOME OF YOUR WORK FOR YOU.

IF YOU'VE USED IT TO AUTO-COMPLETE A SENTENCE—MAYBE WISHING YOUR COWORKER A GOOD WEEKEND—YOU'VE USED THIS TECHNOLOGY TO WRITE.

AI MIGHT HAVE SLIGHTLY CHANGED WHAT YOU SAID—MADE IT MORE FORMAL, OR MORE POLITE.

LOCAL TECHNOLOGY AND PUBLISHING EXPERTS ARE NOW SAYING THAT THESE SMALL CHANGES WE MAKE USING AI ARE SHAPING HOW WE WRITE AND COMMUNICATE—IN WAYS BEYOND JUST EMAIL.

ANAMNESIS

**LITERATURE—BOOKS—ARE NOW BEING
WRITTEN WITH THE HELP OF AI.**

WRITERS SAY THAT

THEY AREN'T
THREATENED.

BUT SMALL BUSINESS OWNERS AND LOCAL EXECUTIVES ALIKE, HERE IN THE BAY AREA, THINK OTHERWISE.

THEY SAY THAT WORK MADE BY AI SHOWS IMMENSE PROMISE FOR THE FUTURE OF BOTH THE NEWS AND PUBLISHING INDUSTRIES.

TONIGHT KXXX'S RICHARD MARTIN IS GIVING YOU AN INSIDE LOOK AT WHAT THE LOCAL EXPERTS HAVE TO SAY ABOUT WHAT A VIRAL NEW ESSAY MIGHT MEAN FOR THE FUTURE OF WRITING.

TAKE PACKAGE--

(B ROLL: STUDENTS WORKING ON LAP-TOPS IN UNIVERSITY OF CALIFORNIA BERKELEY LIBRARY)

A NEW ESSAY PUBLISHED BY AN AI BOT HAS GONE VIRAL ONLINE.

THE PRESS THAT PUBLISHED IT, INSIDE THE CASTLE, SAYS THAT THE ESSAY WAS CREATED ENTIRELY USING NEW AI TECHNOLOGY.

NO HUMANS WERE INVOLVED IN ITS WRITING AT ALL.

CUE THE SHOCK WHEN THE OVER-WHELMING FLOOD OF REACTIONS POURED IN.

ONE TWITTER USER SAID: "WOW SO CRAZY. MOST PROFOUND THING I'VE READ SINCE I WAS IN COLLEGE. I DON'T USUALLY READ STUFF THIS LONG NOW. EVERYONE CHECK THIS OUT!"

INSIDE THE CASTLE PRESS REP ON ZOOM: 13:42-13:43: "THE OUTPOURING OF INTEREST AND SUPPORT FOR THIS PIECE SHOWS THE KIND OF PROGRESS THE TECHNOLOGY HAS MADE, AND ALSO THAT THERE IS A REAL APPETITE FOR THIS TYPE OF WORK. IT'S A BIT FUNNY THAT THIS WAS THE THING TO GO VIRAL..."

B ROLL: WRITER WORKING AT DESK ELIZABETH, A 35 YEAR OLD WRITER IN OAKLAND, SAYS SHE CAN'T FORESEE A FUTURE WHERE SHE DOESN'T USE AI IN HER WRITING—IT FEELS INEVITABLE.

BUT SHE CAN'T AFFORD

TO COMPETE WITH IT, EITHER.

ELIZABETH: 12:50-12:54: "IF I STICK WITH WRITING AS A CAREER, I THINK USING AI CAN REALLY HELP ME WITH SMALL STUFF. IT'S REALLY USEFUL TO WRITE COPY, WHICH I DO TO PAY THE BILLS."

LEXINGTON DRUTHERS, PRESIDENT OF THE NEWS CORP PRESS OF AMERICA: 6:53-6:59: "HERE IN SILICON VALLEY, WE ARE SEEING IMMENSE PROGRESS WITH USING AI IN OUR LOCAL NEWS OUTLETS. WE ARE USING IT AS A SUPPLEMENTAL TOOL TO WRITE HEADLINES, SHORT CLIPS, AND EVEN SKELETONS OF LONGER ARTICLES. WE'VE BEGUN TO IMPLEMENT IT IN EVERY NEWSROOM TO HELP OUR WRITERS—NOT REPLACE THEM."

LEXINGTON DRUTHERS, PRESIDENT OF THE NEWS CORP PRESS OF AMERICA, SAYS THAT NOT ONLY IS THIS TECHNOLOGY PROMISING TO KEEP THE NEWS BUSINESS AFLOAT–IT ACTUALLY HELPS REPORTERS FOCUS ON MORE IMPORTANT WORK.

IT'S ONE OF THE MANY WAYS AI IS CHANGING OUR WORLD. AND HERE IN SAN FRANCISCO, WE'RE RIGHT AT THE CENTER OF THIS MOMENTOUS CHANGE.

B ROLL: PRESIDENT OF THE CONSOR-TIUM OF AMERICAN PUBLISHERS ROLAND ROBINSON AT HOME IN SAN FRANCISCO'S PRESIDIO NEIGHBOR-HOOD.

THIS VIRAL ESSAY HAS GOTTEN PEOPLE TALKING–AND THAT'S A GOOD THING, SAYS ROLAND ROBINSON, PRESIDENT OF THE CONSORTIUM OF AMERICAN PUBLISHERS.

ROBINSON SAYS THAT THE ESSAY'S VI-RALITY IS EVIDENCE THAT PUBLISHERS ARE MOVING IN THE RIGHT DIRECTION TO ENSURE THE LONGEVITY OF THE BUSINESS.

ROBINSON: 5:17-5:28: "YOU KNOW WITHOUT AI, THERE IS NO WAY THIS BUSINESS CAN STAY AFLOAT. THE NUMBERS JUST DON'T ADD UP. WE SHOULD ALL BE THANKING OUR LUCKY STARS THAT WE'RE ALREADY SEEING SUCH HIGH QUALITY WORK."

M C M A N U S

THE SUBJECT OF THIS VIRAL ESSAY, A POEM TAKING A LOOK AT THE IMPACT OF CROWDSOURCING ON TECHNOL-OGY, IS ALSO A HOT ONE HERE IN THE BAY AREA, WHERE WE'RE HOME TO THE HIGHEST CONCENTRATION OF AI COM-PANIES IN THE WORLD.

THE ESSAY TAKES A CRITICAL LOOK AT HOW THAT POEM REFLECTS OUR CONNECTIVITY: BOTH ONLINE AND OF-FLINE.

ELIZABETH: 13:50-14:14: "I THINK THAT VIRALITY IS NOT PROOF OF SUBSTANCE, OR OF QUALITY. AND PITTING WRITERS AGAINST THE MACHINES LIKE THIS IS A METHOD TO DISLODGE AI AS A USEFUL TOOL FOR WRITERS.

THE ONLY COMPETITION HERE—THE ONLY RACE—IS THE ONE TO THE BOTTOM! CRE-ATED BY BUSINESSPEOPLE WHO DON'T CARE ABOUT WHAT'S GOOD—ONLY WHAT SELLS."

THE ESSAY HAS INDEED SOLD WELL.

RELEASED BEHIND A PAYWALL, MORE THAN 5,000 PEOPLE HAVE PURCHASED IT, AND THOUSANDS MORE HAVE SHARED IT ONLINE.

PORTIONS OF THE ESSAY AND REACTIONS ARE SPREADING ON THE SOCIAL MEDIA PLATFORM TIKTOK AS WELL.

TIKTOK USER GAIAMAMI 0:02-0:16:
(QUOTING THE ESSAY) "DESPITE THE
PROFOUND IMPACT OF TECHNOLOGY
ON OUR LIVES, THE ACTUAL CONNEC-
TIONS – THE WIRES, THE SIGNALS, THE
BINARY CODES – REMAIN UNSEEN."

TIKTOK USER CUBIC_ZRCON1A 0:02-0:09: HEY GUYS EVERYONE SHOULD SERIOUSLY CHECK OUT THIS NEW ESSAY. IT MADE ME FEEL REALLY SAD AND I DON'T KNOW WHY. APPARENTLY IT'S WRITTEN BY AI. I JUST HAVE THIS WEIRD FEELING IN THE PIT OF MY STOMACH NOW. I LINKED TO IT IN THE COMMENTS. I WANT TO KNOW WHAT Y'ALL THINK.

B ROLL: MORE ROLAND ROBINSON AT HOME.

AT 85 YEARS OLD, HE'S SEEN THE BUSINESS GO THROUGH PEAKS AND VALLEYS. AND HE THINKS THAT USING AI CAN INCREASE PRODUCTION, AND MAYBE EVEN BRING MORE BOOKS TO THE WORLD–SOONER, FASTER AND CHEAPER.

ROBINSON 15:23-15:25 "I'VE RETIRED IN SAN FRANCISCO BECAUSE MORE THAN ANYWHERE ELSE IN THE WORLD, PEOPLE HERE I RECOGNIZE THE VALUE OF USING AND PRIORITIZING THE LATEST TECHNOLOGIES."

B-ROLL: PHOTOS OF MARILYNNE (AT HOME, IN THE LIBRARY, AND WITH HER SCULPTURES): MARILYNNE, WHO IS A 72 YEAR OLD LIBRARIAN, IS THRILLED TO STILL BE WORKING. SHE'S ALSO AN ARTIST WHO LOVES SHARING HER ART. BUT SHE SAYS SHE WON'T BE ABLE TO AFFORD LIVING IN SAN FRANCISCO IF SHE STOPS WORKING—THE COST OF LIVING HERE IS JUST TOO HIGH.

MARILYNNE: 12:25-12:30: "I AM CONSTANTLY WORRIED ABOUT LOSING MY JOB. MY HOURS HAVE ALREADY BEEN CUT IN HALF IN THE PAST TEN YEARS. I SEE WHY PEOPLE ARE EXCITED ABOUT THE NEW TECHNOLOGIES AND SOCIAL MEDIA, BUT THIS IS THE REALITY OF OUR WORLD—EVERYTHING IS MOVING ONLINE AND INSIDE."

THE SAN FRANCISCO CITY BUDGET FACES MORE CUTS IN 2024 SINCE TECH COMPANIES HAVE FLED THE CITY FOR CHEAPER REAL ESTATE AND CLEANER STREETS.

EVEN THOUGH THE CITY STILL HOUSES THE GREATEST NUMBER OF TECH OFFICES IN ANY CITY IN AMERICA, RISING CRIME RATES AND HOMELESSNESS HAVE SCARED SOME BUSINESSES OFF.

THIS MEANS THAT THE CITY FACES TOUGH QUESTIONS ABOUT HOW TO MAKE THE NUMBERS WORK. BUT ONE THING IS CLEAR: SOMETHING'S GOTTA GIVE. AND THE PROMISE OF AI–A BOON FOR DOWNTOWN–COULD HELP SOFTEN THE BUDGETARY BLOW.

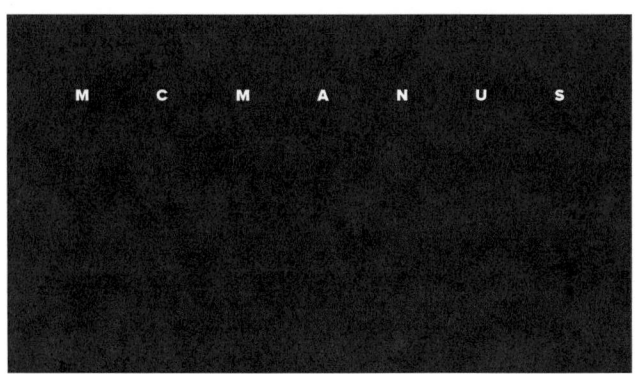

B ROLL: SF MAYOR LONDON BRIDGE WALKS WITH RICHARD MARTIN DOWN MARKET STREET IN SAN FRANCISCO.

BRIDGE 17:25-12:30 "I THINK THAT THE POTENTIAL FOR AI NOT JUST TO HELP DEVELOP NEW TECHNOLOGIES–BUT REVITALIZE OLD INDUSTRIES LIKE PUB-LISHING–IS ONE OF THE MOST EXCITING PARTS ABOUT HAVING THIS INNOVATION RIGHT HERE IN OUR CITY."

ACROSS THE BRIDGE IN THE EAST BAY, STUDENTS AT UC BERKELEY WERE NOT SO OPTIMISTIC. A GROUP OF STUDENTS HELD A SMALL PROTEST AT HISTORIC SPROUL PLAZA TO SEND A MESSAGE:

B ROLL: 10 STUDENTS CIRCLING THE PLAZA WITH SIGNS. NAT SOUND: "HEY, HEY, HEY HO HO AI WRITING HAS GOT TO GO..."

STUDENTS GATHERED TO BRING AWARENESS TO WHAT THEY SAY ARE UNFAIR LABOR PRACTICES IN THE THIRD WORLD THAT HELP AI RUN. CHANCELLOR BUTIMA CHIER-LOEDER OF UC BERKELEY SAYS THAT THE CONCERNS OF THOSE FEW ARE FAR-FETCHED.

SHE SAYS THAT ANY WRITING DONE BY AI AT THE UNIVERSITY MUST BE WRITTEN WITH APPROVED ALGORITHMS— AND THAT THESE ARE ALL UP TO THE HIGHEST INDUSTRY ETHICS.

CHIER-LOEDER 17:33 - 17:48: "WITH THE GUIDANCE OF OUR INDUSTRY PART-NERS, WE HAVE SET THE LEADING AC-ADEMIC STANDARDS FOR AI WRITING. WE ARE PROUD TO OFFER PROGRAMS SPECIALIZING IN THIS NEW FRONTIER OF TECHNOLOGY AND ART.

THE POPULARITY OF THIS NEW ESSAY HAS EVEN SPARKED NEW INTEREST IN OUR MASTER'S PROGRAMS AND IN OUR NEW TECHNOLOGY MENTORSHIP PRO-GRAM, WHERE WE PAIR ENGLISH UNDER-GRADUATE MAJORS IN THEIR SENIOR YEAR WITH LOCAL TECH FOUNDERS TO TEACH THEM ABOUT AI AND PRACTICAL SKILLS."

MARTIN, WALKING DOWN TELEGRAPH AVENUE IN BERKELEY.

THE FUTURE OF AI LOOKS BRIGHT. IT COULD CHANGE THE WAY WE READ AND WRITE, AND THE BUSINESS OF BOOKS. BUT IN THE HERE AND NOW, THIS ESSAY SURE HAS SET OFF A FIRESTORM HERE IN THE BAY.

IN BERKELEY, I'M RICHARD

MARTIN, KXXX NEWS.

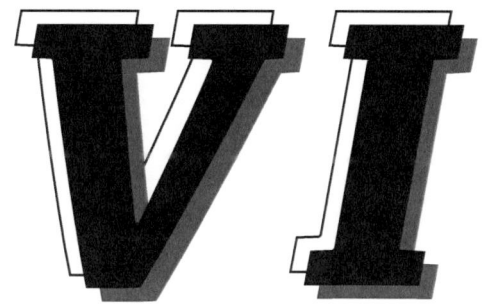

INDIVIDUATION

The question and quest of individuation in Jungian psychology is the highest ideal: it is the task of self-actualization through the integration of the **conscious and unconscious mind**, allowing a person to live wholly as themselves, able to pursue a connected and purposeful life. In philosophy, individuation distinguishes one thing from another—that I am different from that couch over there—delineating sameness from different-ness.

This section addresses the question: among the chorus of technologically mediated, fabricated and crowdsourced voices, who gets the last word? And can the "last word" be considered separate from any other part, since each element of the project is layered on top of the other, all inextricably intertwined?

To conclude, I've taken an informal poll, by phoning up **AI experts**, **artists**, and **friends**. I've conducted brief interviews and transcribed them using Otter AI. The interviews are slightly edited from the Otter AI transcription to ensure that they're mostly legible (though not up to the fact-checking standard surely, and may contain slight mistakes or mistranslations).

Their answers reflect their perspective

on the technology, the project, their relationship to me, their mood, and how I ask
and frame the questions. Their responses will serve as a record for how people
on that day, at that time, thought about AI
in relation to themselves and the world.

M C M A N U S

Paris Marx

Me: Could you also just introduce yourself, or how you would like to characterize yourself?

PM: Sure. I'm Parris Marx, I write a lot about technology from a critical perspective. I host a podcast called Tech Won't Save us. And I write Disconnect, a critical technology newsletter. I also had a book published last year called *Road to Nowhere: What Silicon Valley Gets Wrong About the Future of Transportation*.

Me: Awesome. So what I'd love to hear about from you is kind of like who you think conversations about AI are led by usually? And then how narratives about technology become concretized?

PM: Yeah, so it's led by rich white guys in Silicon Valley. Like, I'm sure there's some women on the teams as well. But let's be real, the tech industry is very male dominated. There are people who often come from wealth, or at least have some degree of privilege that allows them to go through this startup process without having to risk, you know, financial ruin if things don't work out . So the narrative that we have about AI right now, and how the public is understanding it is largely shaped by people like Sam Altman, Elon Musk, that guy at DeepMind, whose name I'm blanking on at the moment, but you know, these really influential figures within Silicon Valley, who have particular I think it's fair to say ideological beliefs around what AI is and how it works.

And so over the past year, in particular, since the launch of ChatGPT, which was a year ago yesterday, I believe. There has been a lot said about what AI is and what it means and what it does. And those understandings of the tool and the narrative around, you know, what effect it's going to have on society have been very much shaped by these influential figures. And I would say, why Sam Altman, in particular, because he is the head of open AI, which made ChatGPT, which was what effectively kicked off this AI hype cycle.

And I think that the issue there is that by going off the narratives that these people have narratives that are really based in this idea that AI is sure an opportunity in some ways, but also this like threat to humanity that could turn into this artificial general intelligence and, you know, oppress us or kill us or turn us into paperclips or whatever. That shapes the conversation in a way that is very unhelpful. And that leads us to have discussions that are not rooted in the actual problems that the technology presents. And that would need to be addressed if we wanted to think about what an ethical use of this technology looks like. Or if there is an ethical use of the technology.

Me: That makes so much sense. And then before getting into sort of the use of the technology, I'm also curious about how our understanding of the production of the technology is also influenced by, as you said, people who have a lot of power.

PM: Totally. I think it's interesting, because, you know, if you think about the Sam Altmans of the world, and when they talk about AI, they would present it as though there's this computer, or, you know, this, this group of computers, this data center, or whatever, that is just doing these magical things, right? It's just turning out this writing that a human could do, or is making art in a way a human could do. And we can hardly tell the difference anymore. And the computers are replacing the humans, and what is this going to mean for our society? Blah, blah, blah, right? And like, for me, I've been at this long enough that I remember hearing all those things like 10 years ago, not even not even 10 years ago, right?

When the robots and they were going to take our jobs, and we were going to need a basic income and blah, blah, blah. So that is the narrative that I hear from these powerful people who are not affected by any of the everyday concerns that the vast majority of people in our world have to consider.

Me: Right.

PM: I think one thing that is notable about this wave of AI hype that's happened over the past year, is that the voices of those workers and of the humans behind the technology have been present in a way that I think has been more rare in past hype cycles. You know, if you think back to like the early sharing economy days the idea that We would listen to like what a Uber driver has to say is like hard to imagine, like sure that came along later, when the negative effects of the model became clear, but in the moment when it was taking off, and when the hype was very present the idea, I think that people were really listening and caring what the worker said, was not really they are like, or in early gig economy, you had like a lot of stories where journalists will be like, go out and drive the car for a day and be like, Wow, this is so great. It's offering so much flexibility and blah, blah, right? But in this cycle, like very early on, you had Billy Perego at Time Magazine, right about the workers in Kenya, who were doing the content moderation work for open AI on ChatGPT. And the really negative effects on them of having to read a lot of the most vile content on the internet. You know, I won't name the types of types of stuff because I don't think it needs to be named, but people can imagine what kind of shit is out there. And since GPT is trained on just all this data that scraped off the internet that naturally goes into the model.

And the company knows that, you know, if you or I, or some random person who doesn't understand the tech kind of goes in and types something, types a little prompt, and it starts spitting out beastiality, or gruesome, like, whatever else, murder fantasies or whatever, that's probably not going to look very great. And it's going to be picked up by the media, and there's going to be a ton of negative stories. So they want to train those things out of the model. But they need people to do that. And so they choose the very poorly paid people, I believe it's less than $2 an hour, in some cases, in Kenya, who do not get mental health support, and who just get thrown out when what they need them for is over.

And of course, the Wall Street Journal and other media organizations have gone back to Kenya throughout the year to speak to these people after a certain period of time, and to hear what the effect of doing that work is hard on them. And so I think, in previous cycles, who would not have heard those voices. We would not have said, Hey, are there people over in Kenya or even other parts of the world doing this kind of work? And what impact is it having on them? And how does it puncture the bubble, or the myth, that these AI CEOs are creating? Yeah, so I think that it has been kind of distinct in that spirit.

Me: Yeah, I really appreciate you putting it into context to about like, the ebb and flow of these kinds of hype cycles, as you call them, too, because it is that and I think it's interesting to to think about, like, how inclusion is perceived in a hype cycle or in like a media cycle. And as you say, right now, there's more receptiveness to it. Overall, say, outside of this news cycle, or even in the past year that we're in, how would you say inclusion is perceived in AI reporting and in AI media?

PM: I would say that, it's generally not a huge focus. And that's not to say that there's never reporting on it, like, I believe it was MIT Tech review did a good series last year, the year before, where they looked into the broader impacts in the human side of AI. Basically looking at this large workforce that is very poorly paid, that goes into creating these tools, and that is largely distributed around the world. It's not just Kenyans in a call center environment. Like there's many different kinds of forms that this takes. And in many cases, the tech industry, and their AI applications are very heavily dependent on that. So we do get it at times. It does surface.

But I think in general, the general frame that we get on it is basically what the AI companies want us to hear, and want us to know, and want us to be talking about. And obviously, I think that's a real problem, because it leads to misunderstandings about what the technology is or can do. I think that it has moderated a certain degree where we're now a year into this.

But if you think back to the early hype around ChatGPT, like the kind of stuff that the media was writing was just kind of wild, right? When you think about how uninformed it was, or even actively misleading the public as to what was going on and what these tools were capable of– the real suggestion that they were thinking like human being, and the ascribing of human-like characteristics to these computer programs really leads people to see them in a way that I think is not reflective of their actual capabilities. So I think it does emerge at moments. But I think it's important to understand the context that this is an industry that is not very diverse, and an industry that is shaped by particular perspectives on the world that are not reflective of the vast majority of the population. And that naturally affects the conversations that we have about it.

Me: Yeah, absolutely. It definitely does. And I want to pull out there when you said people with certain perspectives, because I think that leads pretty much into my next question, where I'm curious, who do you think gets to say and decide which types of work or which parts of work should be automated, and which should remain with humans?

Because I think sometimes that is part of the marketing of these AI products. And it's not even necessarily like these things are happening, but it's what the companies are putting out there like this can change X type of work, this should change X type of work. Curious for your perspective.

PM: Yeah, well, it's the CEOs, right? That's quite clear. If you look back to the last AI cycle, which was in the mid 2010s, right? That was a moment when the tech industry was promising us, promising us that self-driving cars were right around the corner, and they were coming, right. They were also saying that, yes, there were these AI tools that were going to change the way that people write, that we're going to take away the jobs of a lot of journalists, because AI was just going to be able to do their reporting for them somehow. And also, that we were going to have this big rollout of industrial and service automation, so that people doing manufacturing, we're going to have less work, but also people flipping burgers at McDonald's, were gonna lose their jobs and all this kind of stuff, right?

The idea was that we were moving into this new phase of technological development or technological capability that was going to threaten the work of so many of the regular jobs that people depend on. And then you fast forward a few years, and you see that none of that happened. You know, it'd be funny to think, like, during the pandemic, when we were experiencing all these supply chain shocks that a few years before, the tech industry was probably not like promising that like, we wouldn't need truck drivers anymore to transport our goods, then it's like, yeah, that would have been nice right now, probably because, you know, there's this big shortage, and everything's messed up. But yeah, that didn't happen. We still needed the humans, even today, the self driving cars are not here. And they've just experienced another major setback with the incident that happened with crews in San Francisco.

And so, you know, time and again, they make these promises, they tell us what the impact of their technologies on work are going to be. And then often those proved to be lies, right? Because if we look back at the mid 2010s, the impact was not to wipe out all of this work. But to use that narrative to provide cover for companies like Amazon and Uber and later, many others who followed in their footsteps, to roll out technical systems, algorithmic management, workplace surveillance systems, that were able to further reduce the power of workers, in some cases carved the mode of employment relations all together, when you look at a company like Uber, or to ensure that you transformed unionized well paid work into non union poorly paid work in the case of Amazon warehouses, right. Yeah, and there's a bunch more I could say about that. But I don't think it's particularly relevant to the question.

And so if we fast forward to what's happening today, you know, the argument that we have from the AI companies is basically that, on the one hand, these tools will eliminate a lot of creative work without considering whether that's something a human society would actually want to get rid of, and hand over to machines.

You know, there was a funny comic circulating a few months ago, that it was like, they promised us unlimited leisure, and that the robots would do all of like, the menial jobs, and now they're automating the creative work and leaving, like menial work to the humans. And it's like, yeah, it's kind of the opposite. It's not actually happening, by the way. AI is not going to wipe out human creative work. I'm very skeptical of that.

But then you have these other promises from people like Sam Altman that like it's gonna get rid of law clerks, which is again, something we had heard in the mid 2010s. That it's going to become an assistant for all forms of professions, and so it's going to be able to be your teacher and your doctor. And whatever profession you're in, it will be able to help you do your job. Maybe replacing an assistant if you already have one, but more so like adding an assistant where, where maybe you didn't have one in the path. And again, I'm skeptical of all these things, right? I don't think I don't think we're seeing the technology play out as the CEOs telephones. But then if you think about kind of who has the power to decide how that technology is affecting the workplace, it's not the workers who can look at it and say,

Okay, how could we use new technologies in order to transform, you know, the workplace, the production, environment, whatever, to make it more efficient to make it work better for us, because the goal of you know, technology produced by capitalism, going well, back to the Industrial Revolution, I would argue this really hasn't changed is to provide more control and more power to the employer, right to the capitalist. And even though the form of technology has changed somewhat, you know, we're not talking about factory machinery and stuff like that. We're talking about digital technologies. But I think that through line is still there. So ultimately, it's still the bosses making technology that serve their interest, above the interests of everybody else.

PM: Yeah, absolutely. Thank you so much. There's so much to think about there. I tend to agree with you on pretty much everything you said there. So I'm actually wondering, what do you then think is the future of work? It can be kind of like a funny, almost like, like, trope-y type question. But I am curious if you have thoughts there.

PM: I think it's, I think it's difficult to say like, I think I think about how back in like the 1930s,

I believe it was John Maynard Keynes, you know, who was obviously very influential economist was writing that, you know, within however many number of years, I think it was the next century or something I can't remember off the top of my head. But like, people would essentially be working like hardly any hours a week, I think, like 10 or 15 hours a week. And, you know, there'd be this great like, kind of leisure society and blah, blah, blah. And now, like what we've seen in the past decade is like working hours getting longer again, and you know, people are being paid less for their work, for the most part, you know. And so like these predictions, I think they're nice, and I think they give you something to aspire to. But I don't know, it's hard to put my faith in them anymore, when I've seen how things have actually worked out.

And so I think I would say that the future of work really depends on who has the power to decide who shapes the future of work, right? While that power is in the hands of, you know, AI CEOs are tech CEOs or capitalist more broadly, you know, they are going to shape the future of work to their interest.

And they would often tell us that that means automating work, and then you know, how are people going to get paid or whatnot. But I think we actually underestimate the degree to which they need human labor in order to power their business models and their technologies and things like that. Their goal is not to eliminate human labor, but to hide the human labor and devalue the human labor and disempower the human labor. And that is what the goal of many of these technologies ultimately is when you dig into them. And, you know, it's in practice what many of these companies have done with the technical systems that they've created.

I think that my worry is that the longer it takes for us to wake up to that reality from the tech companies and to challenge the power that they have, the further down the road, we're going to get to having this totalizing technical system that has a significant degree of power over us. And that makes it harder and harder for us to push back against it, and continues to transform work in such a way that people are paid less for their work, there are higher and higher production targets placed on people, there's greater precarity in terms of how that work is done. So maybe you won't have a stable job anymore. Maybe you'll be looking for gigs, and they'll be flexible hour contracts, then who knows what that's going to look like for you.

Of course, we know many people have been pushed into that type of work over the past decade. You know, I think that there is obviously a future where people have more power over technology and over the workplace and over what the future of work looks like. But the power needs to be built in order to achieve that future and achieve that power over what work is going to look like right? You know, we didn't get a 40 hour work week, an eight hour day just because of the generosity of bosses. It wasn't because it was won by capital. It was won by workers.

And I think one of the hopeful things about this AI hype period as well, is that we have seen a lot of unions and a lot of workers being very vocal about what this technology means to them, and not being willing to buy into the narratives of the CEOs and the companies.

Most notably, of course, was what we saw with the writer's strike and the actor strike in Hollywood, where, unlike a lot of creative workers that don't have that kind of collective body that negotiates directly with the companies that are responsible for getting their creative product out into the world. Like, do you think book authors, you know, they don't have a kind of collective union that negotiates with the publishers to decide if AI is okay to use in publishing or whatever.

But in the case of the film industry, because of the, you know, very long legacy because of fights that are, you know, a century ago, or more than that now, that resulted in the unionization of the industry, they were able to go to their, you know, collective bargaining this year. And demand that restrictions on AI were put in the contract to try to protect their workers in their industry and their form of work. In part because of how they've seen how digital technologies have already been used against them with the streaming services and things like that. And so I think that was hopeful.

And I think we saw many other kinds of professions or unions, or what have you. Also look at what these technologies might mean for them. In some cases, you know, they looked to have language and collective agreements. In other cases, they simply, you know, organized petitions, organized protests, you know, tried to draw attention to it through the public. And so I think that all those things are positive. Does it mean that, you know, we're going to that, that workers are going to control the technology anytime soon? You know, probably not. But I think it's a step in the right direction. And I think it helps to show people that, you know, the only leverage over technologies and from a user, which actually has very little leverage, or from just the government in terms of its regulatory power, but there's another site of power, if it can be developed and renewed, to challenge. Yeah, I guess what these companies do and what their vision for how these technologies should work, is?

Me: Yeah, absolutely. And I appreciate you characterizing that even as a third site of power, because I think that there's a narrative from tech companies, and even just kind of like, out there in the world that oh, it's either you as a user, or it's the government regulator, as binary?

Well, you know, those are really all of my questions. And I really appreciate all of your thoughtful responses. If there's anything else you would like to add on anything else we were talking about, I'm happy to.

PM: I think probably the only other point I genuinely make about AI, and it touches on something that you asked is really, you know, the question of who decides which work ultimately gets automated, or that AI ultimately gets applied to? Or not even work necessarily, but just which parts of society? One of the critics of technology that has been at that has received attention that he probably hasn't had in a while, during this period is Joseph Weizenbaum, who built the first chatbot back in the 1960s. And who after having done that, you know, he obviously didn't use the word chatbot, back in the 1960s. That's a more recent word, but after building Eliza, which was what he called it. He had this revelatory moment where he saw how people interacted with it. And he became much more skeptical of this technology, and spent the rest of his life warning people about AI and these technologies and humans' relationship to computers. And one of the things that he really argued was that, you know, these technologies are good for calculation, right?

They're created to calculate numbers and

things like that. But they're not good at judgment, because they don't have human values And so, a technology and AI system and AI tool should never be used to make decisions about a human. You know about someone's life. So an AI system should never be decided I think whether you can get a visa or whether you can get welfare benefits, or whether you should be able to keep your job or any of these sorts of things, right, because an AI does not have the values, the human values to be able to make that sort of a decision. Sure it can calculate revenue forecasts, and whatever else, that's fine.

But I think that when we have these conversations, it's important to not just think about what is the technology capable of in terms of, you know, what can it theoretically, or, or technically do, instead of, you know, thinking about the deceptive narratives that come from the sad moments in the Elon Musks of the world. But also, once we understand what it can do, then having the discussion as to what it actually should do in human society. And that we're kind of okay with allowing it to do.

And so I think that comes to decisions about, you know, humans, and whether we're okay with that. And I would say we have a whole load of examples already, where humans have been poorly treated, because of AI systems have been discriminated against, have been denied welfare benefits unfairly and unjustly, and all these sorts of things. And I think that that's a discussion that, you know, the AI CEOs don't want us to have. And it's part of the reason why they mislead us with, you know, the narrative about AI, potentially ending the world or killing all humans are whatever, right? Because it makes it seem much more existential, when the threat is much more grounded in reality, and they don't want to look at that.

And, of course, I would also say, the questioning of not just the judgment part of things and the decisions. But one of the big focuses in this AI hype cycle has been on creativity and automating creativity. And, again, I just think that that, that creativity is something that's inherently human.

And the idea that we would hand it over to machines, just seems so wrong. It's such a misunderstanding of what it is to be human, and what enriches human culture and human society. And that's not to say that people can never use technologies in the creation of art, and we can debate as to whether some uses of technology and the creation of art resulting good art. We still know that people have often used technology to help them in the creative process. But I think that we get into a really serious we get into a place that that's really harmful, and really kind of scary when we start to say, okay, let's just let these things—churn out a bunch of stuff and put it out into the world and dilute the art markets, and the art industry and the jobs that all these people do. Because I don't think it provides the same level of...we just don't get the same degree of enrichment out of it.

And I would just finally say that, I think it relates back to this, I've read a lot of Marxist political theory and economic theory and stuff, right? But Marx, of course, talks about exchange value and use value. And so when we use something, that use value is what we, as humans get out of it. And maybe that can't be fully calculated by the market, and can't be captured in the price system. But when we consume art, when we consume great writing, we get something. It enriches the human part of us, I guess, maybe it would go so far to say soul, I think there are debates on whether we have the soul or not, or what?

Me: I like soul!

PM: Yeah, so it was good. I think even if we're not speaking about the religious side of it, I think it's a good thing to refer back to. But then there's the exchange value, and there's what can be captured by the market. And when I think about this push by AI CEOs to automate creativity, it's part of a recognition that a lot of art and a lot of the benefits of creative work cannot be properly captured by the market.

Because their value is in the use value—in our interaction with it, and what we get from it. And I see the shift to AI art as a further form of commercialization and trying to make art fit into a market system. So you can churn out a lot of things that can be sold, or can be easily commercialized by particular companies in a way that maybe not all human art can.

MCMANUS

Jared Fellows

Me: Alright, who are you?

JF: Hi, I'm Jared fellows. I'm an Angeleno, born and raised. I work in post-sound audio for TV. I like making coffee, roasting coffee, playing fighting games.

Me: Wait, playing what games?

JF: Fighting games like, like Street Fighter, or like, Mortal Kombat...

Me: Is that computer games?

JF: Yeah, I mean, you play them on consoles, too. Have you ever played Street Fighter? Have you ever seen Street Fighter?

Me: No.

JF: Oh, okay, so yeah, they're like 2D games. And say essentially, like a player versus player, in control of these avatars, that have a set of moves. And the goal is to defeat the other opponents, you are hitting them to reduce their life points until they're knocked out.

Me: With real people?

JF: No, these are video games. Yeah, these are video games.

Me: But are you playing against another real person through the video?

JF: Yeah. You can play against, you know, AI players, as well. But the big community around it is like, yeah, player versus player. Cool stuff. Yeah. They are pretty convenient that way. Really fun. Lots of variety in types of games. Yeah, they're really fun. I enjoy them.

Me: Ok, that's cool. I didn't even know you liked those. And I don't know anything about them. I don't know if that will impact your answers at all. But you know, what do you think about AI?

JF: You know, I'm kind of mixed on it. On a very surface level, when it was starting to come out, I know a lot of people were kind of doom and gloom about it. And I think my first reaction is always like, people just react that way to new technology. They think it's gonna be the end of the world. And things won't be the same, which is true, but like in a really bad or remarkable way.

And then I still think that's kind of true. I don't think it's like the end of the world. But I think what's latent in a lot of people's anxieties, too, is that we know the society that we live in. The culture that we live in, and we know how this will be used. Right? I think we're anxious more about that than the technology itself.

I think people are just aware that corporations, that people who have authority, will make choices to, you know, remove their work, or limit them or justify exploitation. In terms of removing the amount of people who can do a job or, or needing a job...it's just the end game of companies not caring about their employees. If they can do without you, they would, and that's just like a sign of that.

I think this is just a result of it. And I know, now, especially, I read an article in NY Times about AI being used in military weapons, which I didn't think about. But yeah, AI making calls on targets, and when to act and stuff like that. That's another really scary thing.

Me: Totally.

JF: So yeah, the application is all I think most people are afraid of.

Me: So you have mixed feelings about it. And you mentioned talking about it in the context of work there as well. Does it come up in your work at all?

JF: Yeah, um, less directly for my specific job and post-production audio. I think a lot of the discussion will have to deal with the recent strikes were about with writers, and them being able to write—like AI being able to write scripts. Or, in the case of actors, AI being able to use their likeness in perpetuity, which is actually even a union critical point, because a lot of union members are the sort of rank and file background actors and people don't think about that.

Those people are paying dues and are feeding the union, if they can't get work, or they're cut out of work because of AI that might, frankly, collapse the union like well, who's gonna pay for it? So there's fear in that way. And I think that's very real. I don't think that's abstract. I think that's actually very like, it's here, it's now this can happen.

For our work, my work specifically the stuff that's real right now, I think, frankly, is attacking work that nobody really cares about.

Me: What is it?

JF: Yeah, it's like foreign dubbing in particular. Like, there is a working program that can change the movement of lips. Like changing the movement of lips on screen. And make it look like Spanish and then use the same actor's voice and just have the words be Spanish.

Me: Oh, wow.

JF: Yeah, that's real, that's here. And that just completely removes foreign dubbing and rerecording, all that kind of stuff. Like there's no need for it. But also like, Do people really like that? Like, you know, to me, it's like, job, it's a work. Do people feel passionately about that. I'm
not sure.

Me: Do they?

JF: I don't think they do. I mean, I'm curious. I might ask, like, I know somebody at work, who does ADR. Because I'd be curious about that. But they do ADR for like...

Me: What's ADR?

JF: That's when they replace or loop lines for various reasons. So it could be like, lack of clarity from a recording on set, or, you know, they want to add a line to increase clarity to a situation, or add drama. You know, for a lot of different reasons. But, basically, like, dialogue has been recorded off set for the purpose of the show or movie. ADR stands for automated dialogue replacement. That's what it's called.

Me: That's so interesting.

JF: Yeah. So yeah, I think a lot of like, the sort of general work, I think, will still be there. But when I can see, that'll affect my work more dramatically later, I think because I can see a lot of stuff. A lot more stuff in the process is being automated or automatic to a point and then having the human touch. And I can see that reducing a lot of the number of people that will be involved in the process.

Me: How do you feel about that?

JF: It's tough. I think that's like, yeah, it's tough. It's, it's, it changes, I think, the nature of like, of the labor and art of it. And that makes sense. Because on one hand, you do have this artistic way of being like, well, you know, a lot of the menial labor of it is done. So we can think more broadly about the ideas, go beyond the presentation of the sound–which is like, what we're really about in the day, right? Like that's, that's really what we're trying to do, service the story and contribute to it. And part of it is the menial tasks. In the same the way that we used to like cut film and stitch it together, and all that stuff that is now done away with. And by doing so, we can actually focus a lot more on the art part of it.

Simultaneously, it is removing a lot of jobs by people who are just workers. But yeah, for people who just treat this as work, I mean, they're just our jobs. And that sucks, too. And I think it also does hurt people coming up to you know, because it's like any apprenticeship, right? You're doing menial stuff, or like the more laborious stuff, right? And as you do that you start to understand and more conceptually, and that's how you kind of go into those bigger, bigger positions of a supervising sound editor, a recording mixer. Because you've done the labor so much.

Me: Yeah, I imagine doing that work would help you conceptualize and understand what you're doing in a deep way, and create a kind of mental framework or structure. And then if you don't have that, it might be different or more difficult. Or maybe you'd even think about the work differently.

JF: Yeah, absolutely. I mean, there's stuff has already happened within post that way. I know like, and dialogue editing, one of the things you do is you remove mouth clicks and pops.You might hear it in raw recordings (clicking sound).

And there are programs now that we use that aren't AI exactly, though sometimes they're blurring the line, that can get away like they can do away with that stuff really fast.

But I know when I was learning dialogue editing, I had a sound supervising sound editor say like, don't use that program. He's film, he's like, do it that way. And his intent was really to show, why we are doing this and have a different sense of hearing as well.

Because if you do it in the the program way, you actually get rid of everything. It sounds kind of bad, right? And the idea was like going back to the basics and really trying to massage and like the harder way, but like the harder way you can get, I guess more of a better understanding of what we're trying to do here. So maybe that can be a similar thing. Like maybe you just do things manually until a certain point like well, I get what I'm listening for automatically.

Me: I appreciate all of it. It sounds like it actually does affect your work a decent amount.

JF: Yeah, it just hasn't actively yet. A lot of this is more conceptual. I mean, the stuff I was talking about, but ADR is here. I don't know how many people have used it yet. But that's a real thing. T

But I think it's not, it's not hard for me to imagine, the artistic framework of this, right? Like, I mean, it's the method in which you do it, I could see a world in which you can prepare a soundtrack to fit a certain template for an AI to understand, and then have it be mixed and do a decent job. And I don't think that's actually hard to imagine.

But if you do that, then you also are working towards working within the confines of frameworks of like a machine to understand. That necessarily changes how you work and how you think about and what you can do. That also changes the art of it. But like I said, that's not yet a real thing. That's just me thinking about like, what's possible.

Me: Right. And I imagine you're among a lot of other people who are also thinking about it similarly.

JF: Yeah, I think the people I'm around are a lot more. They're a lot older, often, especially in post. And so you know, it's funny, because some of them are very doomsday about it. And they see a lot of turnover, right? They worked on tape, and they went through the whole digital transformation of film. And so I'm kind of surprised that they're like, well, this is gonna end everything. But it might be just because they're like, 50 years old. With robots? I don't know. Are you afraid of robots?

CM: Any other thoughts about the project? Anything?

JF: I was talking to Emmett about this. And some of the things we were thinking about was, how there's this whole idea of pop culture being stale, of using the same stuff again and again.

So we were kind of talking about that. And it made me think about, like, you know, AI only knows whatever we necessarily feed it. Now, you know, it's not necessarily generative, totally on its own. And so I was just thinking, like, we were using AI for like scripts, and like other human interaction and stuff like that, like, at some point, it'll just be like, rote and stuck and uninteresting. Temporal contexts. Yeah. Or it can be, theoretically, right. So, yeah, depending on how much we start to rely on it, I'm like, okay, then we're just like it. We're not generating anything new. We're just constantly using AI stuff. We're actually kind of in a rut.

Me: It'll be boring.

JF: Yeah, boring, uninteresting. Non-contributive. I don't know, it's like, I guess, an extreme version of Hollywood telling the same stories again and again, or we are remaking the same stories again, and again, like having no new ideas. They will literally not, because they're just spinning whatever was like current at the time of creation.

Me: Yeah.

JF: Yeah. And also something. Final thing. You know, Blade Runner? 2049. Yes. So I remember when I watched that, I liked the one I still

Me: Wait, that one has Ryan Gosling. Right?

JF: Right, yeah.

That one. But one of the things I liked about it, at the time was, I thought it you know, because Blade Runner is, is very much kind of pushing the buttons in the question of like, what really makes us human and keeps kind of complicated, like, well, robots have this robot said this, like, what is where's the line, and it keeps getting blurred? Or more difficult. But I thought it did an interesting job with Dawson's AI girlfriend in there. And I remember the time feeling like the movie positively affirmed, like yeah, like these relationships we have programs are real, even if it's like, not a real person, our interaction emotions and the way we engage with it is like very real and valuable.

And I feel very positive about that. And this was in 2017 or so like that. And so I was just thinking about that again, I was like, you know, I wonder how I feel about that now, if I watched that now. Is that what the movie was actually trying to say? Would I change my mind now that AI is an actual real thing? I could actually have an AI girlfriend in my life probably in my lifetime if I wanted to. I don't think that's far away now.

Me: Do you want an AI girlfriend?

JF: I would take one just to try. Just to say I've done it.

Both: Laughing.

MCMANUS

Diana Azevedo-McCaffrey

Me: Alright. Who are you?

DAM: Diana Azevedo-McCaffrey.

Me: Nice name. How would you identify yourself?

DAM: Um, are you looking for pronouns? Or what are you...

Me: There's no rubric for it. I'm being really vague too. Maybe you can say your job, you can say how you want to be identified. What you like to do, however you want to be identified.

DAM: So I'm a Policy Analyst Center on Budget and Policy Priorities. Yeah, I guess so. Stick with that.

Me: Do you want to say what you analyze?

DAM: Sure. So I work on anti-poverty programs. And I specialize in direct cash benefits. And also my areas of interest in concentration are anti-poverty issues that impact women disproportionately. So that can range from domestic violence or reproductive rights and access issues. And trying to figure out how direct cash benefits can help alleviate poverty through a race and gender lens.

Me: Amazing. Okay, so what do you think about AI?

DAM: Um, I guess I feel like you know, I wish I had prepared for this...

Me: No, no, it's just what you think.

DAM: I feel like it definitely presents some challenges. Are you looking for just like generally, or kind of within a certain context generally?

Me: I'll ask you about how it impacts your life, if it does at all. And then how it impacts your work, if it does at all.

DAM: I think I guess within my work I don't so far like we haven't seen I guess like direct impacts or knowledge. I don't think that governments are usually on the forefront of technology. To my knowledge, we haven't seen it used within government programs.

What worries me about AI is maybe not

necessarily AI itself, but who has the power to use it and how it may perpetuate inequities and injustice. I think some of it is like in terms of the rhetoric and the headlines, you know, that we see around like, oh, AI is gonna be AI is like, coming for your jobs and like taking over everything and stuff like that. I think that rhetoric is just blown out a bit of proportion.

I think I feel like it could be somewhat seen as maybe a next frontier. I don't know, like, the industrial revolution or one of those times in history. So I think what worries me about it, I guess within that type of historical context, like if you're thinking about it in that framework, is how it will be harnessed to, you know, make inequalities worse, and kind of just perpetuate the problems that we already have. Just because that's where kind of the money lies. And that's where, like, personal interests lie. So I think I lack the technical knowledge about AI, but it does worry me in that context...

Me: I feel like even you saying you lack the technical knowledge, which, sure I do, too. Maybe a lot of people do. Even that, or the idea that someone can't have an opinion about it, or doesn't know enough to form an opinion about it, if they aren't sufficiently technical. Even that is funny to me, actually.

DAM: Yeah, that's true. That hat does say a lot about the exclusivity of it. In that , who can harness AI, and for what?

Me: And who gets to have an opinion about it?

 DAM: Yeah, who gets to have an opinion about it? Definitely. Like, so far, I feel like with AI, like, within, with, like, how I've been exposed to it. It's like, in a way it's like, like, I feel like I'm probably being impacted in ways that I don't really notice. We all are.

And then, but then there's sort of, like, the superficial aspects of AI that we see, people using it to make pictures and stuff like, which is just silly. So then it does beg the question like, how else is it being used in ways that I'm not aware of?

Me: I'm actually going to do a deep fake of this conversation using your voice after we finish.

DAM: Oh great, okay.

Also, I mean, kind of going back to what we're talking about, about who gets to have an opinion about it. It is true that there are very few women in tech, and there are very few women working in AI.

Me: I know that is true. And I have always found that interesting and disturbing.

DAM: Yeah, I think that is pretty telling, I guess from working, you know, in most of my career has been nonprofits and either like, more like, you know, gender specific issues or poverty issues now. Most of the people I work with, they're like, pretty like diverse workplaces, generally, but they skew mostly women. And so that does that it is worrisome that, you know, we are not getting that diversity or like perspectives. And it does put into question whose interest is AI serving. You know?
Me: And how can you even know?

DAM: Yeah, I was gonna say something, I guess something that I thought of was that is kind of something about AI that I am concerned about. Another thing is using AI for misinformation and propaganda.You know, with, you mentioned, deep fakes, and just all the other things that AI can construct. I think that is super concerning.

Me: Yeah, it's really scary. And it's impossible to know. I think a lot with image and tech stuff to know if it was made by AI now. And also, even funnier, it's impossible to tell what propaganda is, or like the opinion of what propaganda is, in 2023. It's not very specific, I would say also.

DAM: You mean, like, could you expand on that...

Me: Well is propaganda unbalanced reporting? Is it public relations? Is it a combination of these things? Like, right? Yeah, not everyone looks at one thing now points at it and goes, that's propaganda. I think that's kind of funny, too.

DAM: Right. There's a big spectrum now. And there's like, different varying levels of like
internet literacy, as that kind of presents other additional problems. Y

Me: Yeah, I mean, this is totally a big part of my project, because it's like, I recorded three hours of the TikTok spirituality, ghost, alien and stuff that veers into conspiracy.

And I watched three hours of it, you know, like, straight up and was recording and re-corded the text from it.

And like, I don't know, for a ghost story, I don't know if someone's telling the truth or not. But like, you could just, some of it's be-lievable, some of it's not. There's such huge entertainment value in it to that, like, I don't know it. I don't know what the impact of it is. Or what it would be, but I'll tell you, yeah, watching three hours of it started to feel weird.

DAM: Yeah, yeah. Yeah, that's a lot of a lot of time to be like, sucked into that.

Me: I might have been watching propagan-da.

DAM: And it's gonna Yeah, impact people differently.

Me: Yeah, totally. Because I feel like it's, it's almost like too easy for people to now say like, oh, you know, only like, certain type of person is susceptible to this stuff. This stuff like, what do you mean this stuff?

Misinformation or propaganda or things

which are dubious, like, it almost feels like there's no start or end to it. And it's not one type of person who might be open to it or susceptible to it, because it's so broad too.

DAM: Yeah, and I think that I think that these technologies are impacting us in ways that like in kind of, like incremental. They are subconscious. And it's like, how do you qualify that? Or like, how do you identify that type of impact?

Me: Totally agree. Any feelings about that? Or any thoughts on that?

DAM: I do think it's scary, in the sense of, when you're thinking about your will and your agency. And I think for me, maintaining those things is important. I think for a lot of people they are. But it's scary to think about the fact that there are people and compa- nies whose work is very much invested in trying to sell people things. It does put into question one's agency and autonomy, over their feelings, or their thoughts. What do they decide to do day-to-day or kind of, like, with their life?

There have always been forces that impact-

ed us in terms of the norms and expectations that society has for people. But I do
feel like this is a whole other thing. I don't
know if I'm making any sense...

Me: You're making total sense! And I think, I
mean, it almost sounds like a kind of Boom-
er thing to say...but I feel like we're all way
too comfortable with advertising. And we're
like, swimming in it. I think that has a weird
impact too. And all of that is obviously exac-
erbated by AI and algorithmic targeting.

—

Julia Moore

Me: Who are you?

JM: Julia Moore?

Me: How do you identify yourself?

JM: She/her.

Me: Everyone keeps saying pronouns actually. But I meant more like, how do you characterize yourself? Either things you like to
do, your job, where you live...anything like
that.

JM: So I live in the San Francisco Bay Area, born and raised. I am a biostatistician. I have been doing research in clinical and translational research studies for 10 years, oh, my goodness. I am adjacent to an arts scene and have many creative friends. I don't know if you want to put that part in. But I enjoy cultural critiques, and thinking about some of the things you're thinking to talk about, like the intersection of technology and culture. And those things are certainly interesting to me as well. I don't really know I'm kind of rambling...

Me: Okay, what the heaven and hell do you think about AI?

JM: What do I think about it? I think that it is a good thing. But I think that there, I do feel like it is somehow widening that inequality gap in America. It just seems like the longer the project of the United States carries on, that gap will continue to get wider just due to the nature of capitalism. I think that the people in the higher echelons of society will continue to benefit. And I think people in the lower economic status will benefit as well, in ways of automating or streamlining certain processes.

Hopefully it'll affect everyone in positive ways. Maybe theoretically, getting better healthcare, you know, tax systems, or just general like systems will be improved. But I think that what it's going to do to the job market is going to be insane.

Lower economic folks are really going to bear the brunt of it, like so many jobs will probably be automated. But I do think that will also affect people who have these more high level complex jobs. Like their jobs can be totally done by AI, maybe moving forward. But with that said, I do wonder if there'll be like, a renaissance of more jobs that were from a bygone era that were like, more akin to, like, less technologically adjacent....I do wonder if it'll actually have a revival of people doing work that requires their hands or, more craft-centered things.

I even just think about, when I've been going on Instagram and looking at the discover page, it's just all photos that are clearly AI generated, and it's just led me to be like, what is even real anymore?

I feel so just like, all this is fake, which is horrible and really scary. But maybe it makes me think that people will kind of take a step back. And kind of I think the internet might lose a lot of credibility, which is horrible and scary, like in the entertainment sphere, or in the cultural reporting sphere. It just seems like with the introduction of this technology, there just needs to be some really like, gold standard of truth moving forward. Like you know, government bodies might need to organize and have like, a way to fact check and like what the fuck are facts anyway, who knows...

But like, even just for my job, like I was looking at a statistical method the other day, like trying to read about how you like the best statistical methods used in a certain scenario. And my internet browser proposed, hey, here's an AI generated response to like what you should do. And I was like, Oh, that's cool. But I was like, wait, I have no idea where this came from! I also don't know if this just pulled all these different languages and phrases together to craft the best response for me.

We were kind of already running up against some of these issues before with like, when you're reading something on Quora, or Reddit or whatever, like you don't know who wrote that, or like, you have to take these things with a grain of salt. But it is scary, like to see some dogmatic language that might insist like, oh, this is factual.

Me: Yeah, it's, the tone of it is so matter-of-fact. That's something that I'm exploring in the project—the manner of speaking, or the manner that information is portrayed in, it lends it a kind of false credibility.

JM: Yeah, no, totally. Yeah, the cultural implications scare me quite a bit. Or even just like, yeah, why kids are learning or not learning or just yeah, that we've been having this crisis of like, false information for a while. And then it just also lets people be a little lazier, probably with understanding things, doing critical thinking and fact checking.

Me: How do you think it impacts your life?

JM: How could it impact my life? I mean, I actually don't think my specific job is in jeopardy, because there's a lot of project management involved. And it can help reduce time with writing programs, which is a part of my job, but, I'm not a full engineer or something. I don't think my job would go away. So I think I have job security.

How will it affect my life? Okay, again, in the positive ways, I actually think it'll affect my life positively in ways that I fully don't know about or understand. Like, I think that, you know, the world is incredibly complex, and like, you know, these economic systems that are in place. Like, I think so much of what's happening culturally, or socially, sometimes we don't even realize how much of it is informed by geopolitics, or the economy.

And like, if there's ways to kind of improve those things, it's like, okay, maybe the economy will become better or something, which then it means that would impact my life productively, like, even just, that would even really affect my job. Like, we're in a hiring freeze, it's stressful. You know, like, I think there's like larger things that it could do to society that have such a big trickle down effect, people don't always realize. So that is good.

But I think like, looking from the bottom up looking at the cultural aspect and how that's affecting everything else. I do think it could lead to lmore of a, like, culture war that we've already been in, or just more confusion around ideas or dissemination of misinformation or like yeah, I don't know. It'll be interesting to see the 2024 election cycle coming up. I think just the headache of the culture wars is the negative impact. And then the unforeseen positives that it might do, like to social systems or economic systems that I'm not even fully I don't fully understand how the technology is being used, but I Um, those are exciting things that again, I don't really know.

Me: Yeah, have you seen anything in your life right now that it's impacted?

JM: I think the most immediate, like, impact I've noticed, is just how people I know who are allowed to use it in jobs or in school, like, they'll talk about how it's really helped cut down or, like, really save them time. And like, it is an incredible technology. And it's like with my mom with, she has to write patient notes, like, she's able to dictate the notes to the AI and it will concisely, you know, reduce notes or like, I know that people use it to write programs code code for them, my cousin, his friends are using it to write essays. So like, I'm immediately hearing about these individual personalized use cases for it. And in some of those are scary, like, even my cousin is funny, or like my mom, we were talking about how you don't even fully understand what the AI is using. And if it's patient level information, like that sensitive information, the system she's been using had been approved by her the company she's working with the private practice she's working at, but like, nonetheless...

Me: I bet lost people would use stuff that you know, isn't quote unquote, approved. Because people use what's available to them. And once it becomes available, you're not always necessarily thinking about what the consequences are. It's like, if it makes your life easier, why not use it?

A N A M N E S I S

M C M A N U S

ANAMNESIS

MCMANUS

Nat Rees

Me: Okay, who are you?

NR: My name is Nat Rees, and I work as the film Exhibition Program Manager at the UC Berkeley Art Museum and Pacific Film Archive.

Me: And what do you think about AI?

NR: In general, or as it relates to film?

Me: Whichever you want to answer first?

NR: Okay. Um, I think that it could be a very interesting tool. I mean, I don't work with AI. And I don't know a lot about it. But I think that I've had a few interesting encounters. So as you know, my partner, Raffie, is on TikTok, and I'm not. But we spent probably close to 45 minutes going through these different face filters, where you give the AI a certain data set, so that it's like, an environment, physical features, and then, you know, maps it onto your face. And there were some really, really cool things that came out of that prompt. So like, it's interesting, on the one hand, how it's doing things that are unexpected.

So like, for instance, when I typed in that I wanted to be wearing jewels, like, if you were more specific in saying, like wearing a jeweled necklace, or something like that, of course, it could do that. But if you're broader than it, has more of a chance to be imaginative, so that it gave me jewels all over my face.

Me: So you looked gorgeous.

NR: Yes, gorgeous, obviously. But it's something that I think is really interesting in terms of creativity, in terms of creating prompts for humans to broaden their imagination. But then, on the other hand, what the AI is fed is human created. So there is a limit to their imaginative potential, because, for instance, there was an AI, I was hearing about that someone was saying, show me a world without roads, or like a world without cars, or highways. And it literally couldn't do it.

Me: Oh yeah, Big Oil getting in there...

NR: Haha, totally, totally. I'm interested in the data sets in terms of like, what's actually possible. But then, like, in terms of the actual film industry, I was just hearing about how people are theorizing that "Wish", that Disney film, is their first attempt to utilize AI on a really large scale to see if it works, which I don't know if this is true.

But someone was saying that the Disney Twitter post of for this film, like from their account, if you zoom into it actually has like weird, uncanny things like it has supposed to be a fork, but it looks like part spoon or something so that it's like, actually a mishmash of imagery that they like seems clear that it was like generated rather than being made.

And I think that it has a lot of potential for generating imaginative possibility. But I do think that in terms of filmmaking, I think that that is an artistic craft. So I think that people can partner with AI, but I don't think that AI should be creating the entire content, because then there really is this kind of more empty or uncanny quality. Or like that sense of like, things being generated in a way where it's mismatched. So I think that's my general feeling: that it can be a really helpful and interesting tool. I'm thinking like I wouldn't necessarily want to see an AI generated film, but maybe I'm wrong. Maybe it would be great.

Me: I don't think you're wrong. You're right.

NR: Like something tells me that we would be able to tell…

Me: Well, I mean, it's interesting because then you're getting into like, talking about aura…

NR: We're talking about aura, we're talking about, like, intrinsic value of, like, a human touch, which I don't even know that I necessarily believe in. Because, well, I mean, I do, but I think it kind of feeds into that idea that, like, we're incredibly special and the top of the food chain for a reason. And it's like, because of our unique intellect, when, like, every other day people are like, Oh, do you know, dolphins can do stuff for fun? Or like octopi? Blah, blah, blah…

Me: It's like, pigs are geniuses…

NR: So I'm so sick of that essentialist narrative. But then, on the other hand, I think that we bring something to, like our species that may be the, like al generated content. Can't or that, like we could tell that it's, like a bit off or uncanny?

Me: I think we can. Yeah.

NR: It's also because I know nothing about it. But I think people are very reactionary, and just scared by AI potential. But I think for people who work in the creative industry, like people are saying, well, it's going to, like, replace everything. And it's going to, you know, put people out of work, when in reality, I think, like Gen Alpha are already cyborgs. I think we need to utilize the technology and work with it instead of fully against it. Even though I'm a Luddite by nature, I still think that because we can't control the fact that new generations are, are more dexterous, and feel more connected to technology then, like human interaction,

Me: iPad babies grown up...

NR: iPad babies.

Me: So how do you come to this? As someone who, you know, ever since I've known you, you're not on IG, social media. I think that affects your perspective, too.

NR: Yeah, because I'm not I'm not an online person, by any means. I live under a rock.

Me: I wouldn't go that far.

NR: Well, literally, people tell me news in a way that's like, you didn't you didn't know this thing happened four days ago. And I'm like, no.

Me: I've never heard of that Disney movie you mentioned but I didn't want to cut you off.

NR: Well, I only heard about it in the context of this, that someone like looked very closely at this image that Disney tweeted and was like, Oh, this is not like a human did not make this.

Me: That' so weird. The Disney movie I remember recently is that, uh, that jazz one? I definitely I cried at that.

NR: Jazz one?

Me: It's like it's called Soul.

NR: I never saw that.

Me: Is this a real movie? Let's see, I'm looking it up. Okay, it is a real film. And I watched it during the pandemic. But that's a bit funny because my whole project is really thinking about not only what it means to have one...

NR: Wait I wanted to hear more...

Me: Yeah, I can tell you...People around the world are data workers working on this technology which becomes AI. They're all putting their understandings or subjective understandings into the world. It pops out these allegedly objective things that all impact us in ways that we can't even see or perceive. But still there. It's an iterative thing. Like, there's got to be some effect to that. And I don't know that it's something that can be measured, it seems like something you might just feel.

NR: Totally. And obviously, I'm someone that completely believes in an energetic field, which informs like really subtle channels in our body and in our field, in terms of how we perceive things interact with others, and I think that why I don't interact with technology a ton in general, I mean, which is not true. Obviously, I'm on my laptop for work, like I have to use a computer but in terms of social media, I really think it has this deadening effect.

Me: Tell me.

NR: I mean, it's just really interesting that you're thinking and writing about soulfulness. Because obviously, what you're posting on the internet can't have all of your like, I don't know intent and can't carry meaning in the way that like expressions or physical touch or like being engaged in conversation can, because everything is so like two dimensional. So I think that I think that technology necessarily lacks soul. So I see AI is being the same in that regard. But it's interesting.

Me: I'm kind of obsessed with the idea that this technology we think is subjective is in fact, the, like, cumulative impact of what so many 1000s and 1000s of people who wrote things online, but then also, real people, in real time, are going through things to make sure that they're correct. It's much more collaborative and much less technical in that sense, then I think that we are led to believe, and I think that there's some impact of that. I don't know what the impact of that is. But that's kind of the motivation behind my project.

NR: Yeah, and I think that like, what that prompted me to think of is one of my favorite films "Her" and there's this really interesting moment in it where Joaquin Phoenix's character is speaking with, you know, the AI, Scarlett Johansson, and in she's saying, What are you up to? And she's like, Oh, I'm actually talking to this AI generated. Oh, my God, what's his name is Alan Johnson...

Me: Alan Watts?

NR: Yes, Alan Watts. So, she's speaking to this AI generated Alan Watson. He's kind of like, whoa, what, how's that happening? And, and she's like, Oh, well, they just fed all of his writings and, you know, everything they knew about him in terms of his biography, and then like, you know, all of these writings about him and now I'm like, you know, speaking to him. And I think that that's something really interesting in terms of like, recreating an incredibly impactful philosopher by putting all of his writing so it is from him, but then how is the you know, the soul isn't intact. So, is it a useful and exciting exercise to be speaking with this AI but with all of These writings could maybe formulate something that is on the same thread that the real Alan Watts could think or could say.

But that is, you know, that is not him. So, yeah, it's like, it's like how, how much? Or like, how profound can that exercise be?

If it's just like a compilation of his work, which, which could be like really useful in terms of like, mining that for data or ideas, but in terms of actually, like, if you're thinking like, oh, I wanted to, like meet him? And this is that then then it? How would that compare? I don't know. I do think it's a really interesting idea. But is it any different than just reading all of his work?

Me: And in a way, reading is such a personal experience and transmission of information, too. It's always deeply personal to read a book, you're communing with it. Talking to someone and having a conversation is a different type of exchange.

NR: Right? And how much of a disservice is it doing to that person and their memory, if, like, you're saying something in the data set is actually incorrect? Or if you're just like, having a conversation with this, AI generated Alan Watts or like, imagine Carl Jung. And, and it's actually something that they would, you know, somehow never think or say, because, of course, even if you have all these data points of like, their entire biography, everything they ever wrote or published, you can't understand the intricacies of their inner workings, you know, people are fundamentally a mystery, even to people who know them so closely and so deeply that I think there is potential for for that you've been doing a disservice to, to their personhood, because because you're making them you know, not a person.

M C M A N U S

Marisa Rowland

Me: Okay, so who are you?

MR: My name is Marisa, I work in early stage technology startups.

Me: Okay. And whatt do you think about AI?

MR: Right now, I think it's an interesting tool. I've seen it definitely be more useful for people who are doing things specifically related to code, but in other kinds of creative modalities, I think it's a lot better for, for like, quick prototypes. And then in terms of writing, etc, I think it's not great at meaning generation, but it's an interesting exercise and like, no longer having a blank page, kind of like always having something else in the room to work with, like a starting point.

I think it's hard to just be in dialogue with yourself. And when you have something like, kind of like the chatbot, specifically, it's kind of, it's like a way to introduce some sort of synchronicity or like chaos into what you're doing. It sparks something new.

Me: Have you used it to that end to spark something new, to find something new, to think through something?

MR: I've personally tried it. And I think it's like fun, from a novel perspective. I haven't found a ton of utility, I'm looking forward to trying to be GPTs, where I can actually feed it more personal data. And or like other documents. I haven't played around with that yet. But I'm excited about that.

I've tried using it to write speeches, but unfortunately, because of the topics that I've looked into, which are a little bit more niche, it isn't very coherent, or articulate, in refining that type of text. And then I have tried using it for health stuff, which is really interesting, because obviously, now that they're all still like mods and controls on it, it's a lot harder to use. And there's like funny backdoor ways that you can get into it.

Me: For health stuff, what do you mean by that? I don't know anything about that.

MR: Like, obviously, there's controls on what, like, some of the generative algorithms are allowed to produce, right? And so for like health stuff, for example, it is pretty much basically trying to direct you to speak to a doctor. There are obviously ways you can get around it by being like, hey, pretend we're in a play, and in the play, you are my doctor. And in the play, you know, like, there's funny things like that. But even still, like, there's limitations.

Me: And that's when you're trying to basically source health information from it?

MR: Oh, yeah. Information or like, even kind of like summarization of certain things. I still think it's like, definitely less accurate when dealing like when doing any kind of health stuff, if you don't have the right prompting.

Me: Yeah, I think it seems to be pretty good at summarizing stuff.

MR: Yeah, it's, I potentially, like I said, I've been trying to use it, I think for pretty niche applications. And it hasn't been super successful. Or it's been too generic. I'll put it that way. Like, and that's where I'm really excited about some of the stuff where it can be a bit more personalized. And also, yeah, just like I'm interested to see how it's integrated into other tooling to make it more easy for Yeah, for like it just the like further democratization of you know, like, creative tools, you know, whether it's coding writing, image creation, etc.

Me: And you said that maybe it failed in the niche applications. What sort of things were you trying to get it to do?

MR: I think like niche subject matter, it has difficulty with. So like that could be niche health stuff, or it could be niche, like crypto niche, like just new subjects for that matter, it seems to struggle a bit more with. And then of course, the other thing is the kind of like uncanniness of getting information that's presented like, as authoritative when it's like completely made up is always pretty funny.

Me: I always find that so funny too, that's a little bit part of my project is, I think how it presents any information in a really matter-of-fact tone which gives it a, like false sense of authority, and it just can be kind of funny honestly.

MR: Yeah, true. Personally because of the way I've interacted with it, it comes back with things that are so generic in a weird way. And kind of reminds me of some of the other stuff you're interested in, related to TikTok and spirituality where like, a lot of the content I get back is so general, that it feels kind of like having your tarot card spread. It's like you make the meaning.

Me: But yeah, anyway, but even that can be useful kinda like what you were saying, it's like, well, if you if it gives you something, and then you do create your own meaning out of it, that is a use...

MR: Completely. And that's where I'm like, I actually think it's really interesting from a kind of like, yeah, democratized tooling kind of perspective, where it's like, okay, like, you have an idea for an image or something, and you don't have to start from zero. Or you have an idea for, or you just want to kind of chat through something. Recently, I was trying to help my sister with naming her company. And it was funny going through all the different name iterations with it. Like, they were terrible names, like, none of them were good. But like it was, it was just, like, kind of funny to get, like, the juices flowing, being like, well, these are all names I will never use.

Me: Right, right. So how do you use it in your life, or what do you think about the role it plays in your life? Has it come up much in your work historically?

MR: In work historically, I mean, let's see, Icame across some of it in college, mainly related to the art world, and it was looking at specifically AI used for surveillance. And then, AI was, of course, like, really big with agriculture. So when I was working for the agricultural technology, startup, and, how you can create kind of, like, just better models, and discovery around what's like going on with your crops.

And gathering data from sensors. But I think the way it's coming up the most right now, at my previous company, Tlon, was, using it for coding help, but also to, obviously thinking a lot about just where does the data come from? Because right now, sure, you need like, a large amount of data to train a model, but then it gets pretty hairy about, like, how your, how your data is being used, how it's being leaked to these models. And then also, like, does the due to these tools have the appropriate data about you to even give you relevant information? And so there's definitely a lot of, I think, thought going into the space where people are thinking about personal data and privacy. And, how does AI fit into that?

ANAMNESIS

M   C   M   A   N   U   S

Patrick Doolittle

Me: Alright, first of all, who are you?

PD: Patrick?

Me: And then tell me, how do you character-ize yourself?

PD: How I characterize myself, like in terms of what...

Me: Some people give their occupation, a couple of people have given pronouns. Some people are saying what they like to do. So maybe some people say where they live. It's entirely up to you.

PD: Okay. I would say I am a writer and mu-sician.

Me: I would go ahead and say that that's ac-curate.

PD: I would say those are my two.

Me: Okay. So what do you think about AI?

PD: What do I think about AI? Well I should have thought of this beforehand...

Me: You know, people keep saying that. But, no! I want to know what you unfiltered, think. Your true thoughts.

PD: Okay. What do I think about it? I think for the most part, I think it's evil. I think it can only be used for good if it's like an assistant to a person. I think well...how deep are we going to get here? Because I do have one point that I really want to make about it.

Me: Well, why don't we go as deep as you want? Honestly, my deepest conversation so far, non technical conversation, as far as friends, has been with Nat. We were talking about souls. So you can go as deep as you want.

PD: Okay, it's not that deep. I mean, I think I don't think I'm the first person to make this point. And maybe I've seen him sort of ambiently. But my main, the main thing, I think about AI, or the main thing that makes me laugh about the AI space, and the kind of industry and culture around AI, which is, of course, a kind of like euphemism for just maybe Silicon Valley as a whole at this point.

It's funny how these AI machines that are

being invented, ChatGPT and like other AI, etc, its kind of quality is assessed by how well it can create a piece of art, or like, a piece of music or a poem. It's like, that's considered to be the ultimate task that a computer can do.

Like, it's the thing against which other a AIs are assessed against each other based on like, sort of, like how beautiful the art or the poem or the music that they've created is.

It's seen as the highest possible end of an AI bot. That it can write a beautiful poem. And yet, these bots are made by total philistines, who have no respect for art at all.

So it's just kind of admission and who for years have been arguing that like, or have the kind of culture of people who argued that like, you know, like STEM is the highest possible like kind of vocation, or like STEM is the best thing to study, STEM is the highest... the world has been completely STEM has become completely STEM like that, you know, engineers are the true artists, etc.

All these kind of like highfalutin ideas about how science and technology and software engineering and all this stuff is like the true frontier of all basically human thinking at this point. But then, by their own admission, like the highest possible thing that their invention can do is make a song. Or like, a beautiful opinion about a book, or, you know what I'm saying?

So it's like, this irony of: these machines are considered good when they can do something that the people that created them consider a lower form of thinking.

Me: And it makes me wonder if they actually think that, or if it's just what we would call classic projection.

PD: Well, it's by their own admission, right? Because it's like we want the best possible thing our thing could do would be to do the thing that we hate the most and think it's for dumb people. Right?

Me: Right.

PD: And it's like, whatever. Not that engineers all actually think like that. Everybody likes music. But in fact, like so much of the culture of Silicon Valley is this kind of weird, philistine preoccupation with art even as it's clearly being had by people who don't really get it, or something. I don't know if any of this is making sense.

But it's like, so it's not to say that like, oh, people who work at Facebook, or who work for Microsoft, or something, like have no respect for art or whatever. But there is this fundamental pride surrounding the idea of being an engineer and the idea that they're at the true frontier. It's like, maybe that's not the true frontier, if your own technology is being assessed, like, according to how closely it can get to a totally different frontier?

Me: Absolutely. And it's interesting to me when you're saying that too, because also, the logic goes, in this sort of, say, we're calling it like Silicon Valley logic goes from, this is the best thing it can do, which is make a beautiful song, or create a beautiful image, and then it turns to production. It's like, okay, how do we scale that? How do we make that a product as well?

PD: Right, right, right. AI is just as kind of like, it just replicates, like what industry does, you know what I mean, quantifies everything and, and tries to quantify things that are not quantifiable so that they can sell stuff back to you.

Me: This is very much related to when I'm working on in the project, which is, so basically, in order for AI to like, process data, it has to be actually labeled by human people who are putting their subjective understandings of the world into like labels. Like this question, is this question positive, negative, and neutral? This all is data that gets aggregated. And it's used to train machine learning models, and so many people's opinions about binary understandings of really subjective stuff is all embedded in that.

And then, at the end of it is one thing, a lot of the time, that feigns as totally objective, or if it's not objective, it's being presented in this, matter of fact, way or tone.

PD: And even when, it's purporting to give a nuanced conclusion about something, it does it in this very strictly structured way that is still fundamentally like black and white and objective. It's very like, well, this is the case. But also, this is the case, it's probably some combination of the two. Even syntactically, it's still, it's still like fitting a circle into a square at the end of the day, even when it tries to give an answer that is aware of and explicit about the fact that there's no conclusive answer for what you're asking it.

Me: Right.

PD: This is my ultimate opinion about AI: It can't offer anything subjective. And the subjective is ultimately the only thing that matters.

Me: I tend to agree with that. And somehow, when you're saying all of this, it's just reminding me of, the tech culture in the Bay Area, too. And I found that the way sometimes people talk and interact in the Bay Area, because I'm sorry, it is a lot of tech monoculture. There are a lot of awesome people doing awesome things in the Bay Area. But it's unfortunately saturated with this tech monoculture where people present themselves in this way too. And people talk in this way too. I always found it funny.

PD: Well, there's like an annoying post... thing...about Silicon Valley too. Like, everything is post everything in this way where it's like you're like transcending fashion and culture and everything. Silicon Valley doesn't realize how much of a subculture it actually is sometimes.

It's Facebook saying well, we're just a forum, we don't have any control over whatever the people say or whatever. But of course, Facebook actually has tons of control. And it's like totally setting the parameters for the way that conversations happen. Their board has opinions, there's a very specific cultural aspect to what they, do to how they manage their platform, et cetera, et cetera, in the same way, it's like the culture of Silicon Valley.

So like, well, we're just wearing Allbirds. And like, we just dress in a black T shirt, or whatever. It's a really put-on kind of neutrality, that it's just like, so not real. It's not real at all. The Silicon Valley is a very specific culture, a very specific kind of person, a very specific affect. It's so far from neutral.

Me: As if neutrality could exist.

PD: Right. And also, by the way, as if, like, there's some kind of, normative valence towards neutrality, as if it could exist, and also as if neutrality is necessarily good...

Me: As if it's a good thing, or as if it's a value. It's like an imagined value.

PD: Yeah, it's a value for people who have never been confronted with a situation that requires them to really look at their values in a hard way, and decide what they are.

It's like a lack of adversity. Just so much to so much of Silicon Valley culture, it's like people who have never... the vibe is so like, I've never faced hardship...

Me: And cafeterias are great because of it! Or, they used to be.

M C M A N U S

Adam Bradley

Me: Okay, could you introduce yourself first?

AB: Yes, I'm Adam Bradley. I'm a professor, assistant professor of philosophy at Lingnan University in Hong Kong. And most of my research is in philosophy of mind. But recently, I've started to do some work in the philosophy of artificial intelligence, in particular, some issues in AI ethics.

Me: Tell me about any of that.

AB: Recently, a lot of people have become very interested in the development, the very rapid development in AI, technology. And I've always had an interest in these sorts of issues, just because as a sort of philosopher of mind, you think in very abstract terms about what it is for something to have a mind or for something to be conscious or something taught, of genuine understanding. There have been debates about this going back almost a century now—several decades—about whether computers could ever be conscious or have genuine understanding, use language, meaningfully, these sorts of things.

And these debates for a long time felt very abstract, because this didn't seem like a realistic possibility to many people that we'd have technology like this. But all of a sudden, with the development of these large language models, like ChatGPT, were sort of staring a lot of these questions about the philosophical questions raised by AI or just staring them directly, and sort of all of a sudden have to try to come up with answers to them. So that's part of the reason why it's such an interesting topic for me, and I think so many other people, and feels like a very important topic, very timely topic.

Me: I'm curious about what you're interested in with the ethics of AI. And maybe this veers into a philosophical territory...that AI data, I'm very interested in the fact that so many people around the world. predominantly in the Global South, working for very low wages, are kind of transmitting or putting their understandings of the world with language into these binary labels. For example, categorizing language as like, positive, negative, and neutral, and then this is all in a data set, then that spits out a piece of, quote unquote, objective information.

I'm not a philosopher, but I feel like there's

got to be some impact of that. And it's very, it just strikes me as something that it's got to seep out, or you've got to feel it, or it's got to have some kind of impact. And I don't know, but that's a lot of the motivation behind what I'm looking at. I don't know if that impacts anything about what you do with ethics, but just thought I'd give a little background there in any case.

AB: I'm interested by these issues as well. I mean, one thing that I think is really fascinating... no one really understands how these systems operate. I mean, there's certain procedures that people develop for training them using techniques like once you've described getting training the systems categorize data, but basically, you do that or one one methodology for doing that is having a human being basically observe their classifications and say, are you doing a good job of classifying things is whatever the case may be good or bad, you know, a photograph that contains nudity or whatever. And so ultimately, the the procedure that the system is learning or developing for for these categorizations is based in some way on human concepts and human feedback, both at the level of the fact that most of the data that the systems are trained on is produced by people like if it's words, for example, but also at the level of categorization.

Not that the systems can't come up with their own sorts of categories. But there's this weird kind of interplay, where a researchers are trying to understand what sorts of concepts these systems are, are developing, actually, the there's a company called a\Anthropic, that's like a sort of competitor to Open AI, that's just published a paper about this that's trying to like look inside of their large language model called Claude, and figure out basically what concepts that it has.

And it turns out that these systems have all sorts of sort of weird kind of alien concepts that don't really resemble what we would conceive of as natural concepts. But at the same time, the strategies we're using to train them on data do employ concepts that are familiar to us. And of course, when we get output out of these systems, we want it to correspond to the concepts that are recognizable to us.

So there's a weird sort of translation thing going on as well, where the system's getting inputs in sort of human language using a human conceptual scheme. It's doing some sorts of internal processing inside of itself, which is alien to us. And then it's producing outputs that are more or less intelligible to us and seem to reflect our, our language, our conceptual scheme. But there's a deep question of how the system's language or conceptual scheme, if you want to call it that, in ours interact with one another. It seems like they can actually be quite different, which raises all kinds of philosophical questions.

Me: Right, and isn't it the case that it might not understand language in the way we do? It's like a set of most likely outputs, not necessarily language?

AB: Yes, although one interesting possibility, which I, myself take pretty seriously that is raised by the development of these systems is that they, that they might actually be revealing something pretty deep about how our own minds work as well.

So there might be a kind of at a high level of description or abstraction at which our brains just performing these sorts of statistical predictions about the next word that we ought to produce. That's actually a lot of work in sort of philosophy and cognitive science that's kind of predicated on a related idea that basically the mind, the brain is in the business of predicting what's about to happen to the creature in the environment, like at a certain level, if you think about it, it doesn't do a lot of good, very natural picture of what the brain does, what the mind does is it just records information that's presented to it like information from the environment, comes in through the senses, and the mind, stores that information and uses it to make plans and things.

But in a certain sense, that's actually kind of an inefficient way of going because anything you're getting information about in the environment has already happened. And isn't something that you can affect any longer. The only thing that matters to you as like a living creature as an animal is what you know, what happens next night, you need information about what's happened to make a decision about what happens next. But in a certain sense, the whole function of the organism is, you know, the next, the next moment or the next space of time. So, from that perspective, it makes a lot of sense to think at a very high level of the mind as just trying to predict what's about to happen. And so there is a kind of way of understanding the human mind where it's not so different from these AI systems and that they're both at a very high level of description engaged in strategies for just trying to predict the best course of action to take in a given a given task.

Me: Oh, that's, that's really interesting. I've never heard about that. So also part of my project is actually continuing off of my article I wrote about why do people on TikTok believe in magic. And the first part of the project, I actually like, I kept that algorithm on TikTok, I kept it screwed up like that. And then I recorded three hours of it and have just the audio of that. And like a record of it, it's really funny and interesting. So that's some information about why I'm gonna ask this next question. Do you think people can be psychic?

AB: Like a good philosopher, I'm gonna say it depends what you mean by that? If, if the question is, I mean, if the story I was just telling actually about sort of the way the mind works, is correct, then there's a sense in which we are all psychics.

Or I think, in the sense that one of the things we're trying to do, for example, in social situations is of course predict other people's behaviors and to predict other people's behaviors, we also have to predict what's going on in their mind. So try to come up with a sort of model of how they're understanding the world. And so like, there is a level at which you can understand a lot of human behavior, especially behavior. conditioned on, you know, social media or whatever, where people are making decisions based on how they expect other people to react to them as a kind of elaborate form of mind reading, or at least at least predicted mind reading. Whether they call that psychic, I guess I'd say I don't think it's literal mind reading, as so much as prediction that's, that's self fulfilling.

So it's not so much that I can look into your mind and know what's there. But I have an idea what's in your mind. And moreover, with technology, you know, starting with language, but developing all the way to Tik-Tok or whatever AI, we develop ever more elaborate means of influencing what's going on in other people's minds, often in very indirect ways. So. So I'd say, literally, I don't believe that we're, we're psychics but but we do have a kind of almost magical power to understand what other people are thinking and to influence what other people are thinking, you know, almost magical in the sense that we ourselves have very limited, a very limited understanding of how we're able to do this, like people.

There's even something kind of faintly paradoxical here, where, at one level, people are very good at understanding and predicting one another's behaviors, like society just wouldn't function if we weren't extremely good at this. But like, if you didn't, there's like, a lot of evidence and social psychology that if you like, ask people like, what is your friend thinking about right now? Or what is this person's attitude towards you? People are terrible at it, you know, they, they think that other people think about them all the time other people never think about you, they think other people like them, other people, you know, don't like you, that sort of thing, or they think people don't like them when they do. So at one level, we're very bad at reading the minds of others, at least in this kind of intellectual way. But like at another kind of more basic, non intellectual level. We're just sort of predator naturally good adults, but we don't really have much insight into even what's going on in our own minds that lets us navigate social situations correctly.

Me: It's interesting that you say we don't have insight into it. That definitely seems true. And it also leads me to, going back into this TikTok spirituality realm—the idea that you can know or feel something, but you don't know why or how. Do you have anything to say about that?

AB: Well, I mean, I'm just making this up but I think actually that phenomenon is why people believe so strongly in the psychic or mind reading or the paranormal, the supernatural. Where it is this thing where on the one hand, we do we have this attunement with other people or animals or things in the world, that is sort of spooky or uncanny. Like, just extremely you know, you can watch sappy videos on, on the internet of people being friends with like, tigers or whatever. And, at one level that makes perfect sense to us, like, you know, the tiger's a big cat, and the person's friends with the cat.

But on another level, it's completely insane that an animal like a human being and an animal like a tiger could have a friendly relationship with one another. Like, we don't have any idea how that works. When you're interacting with, especially something like a nonhuman animal. You know, you don't have any idea what a tiger is thinking or what a dog's thinking. You think you do. But that goes back to the tension, I was pointing out. People are like this with their pets, right? They think they understand what their pets are thinking. And, at one level, that's true, people are very good at manipulating or being manipulated by their pets. But at another level, there's no way a dog understands what I'm thinking, or I understand what the dog is thinking.

It's just literally untrue that either one of us has any real ability to understand the mind of the other. And yet, obviously, humans can have close relationships with animals and of course, the same things are true in a different way with other people. Where, of course, we have some better idea of what's going on and other people's minds. But again, I'd say less, less good than we probably think. And so I think it just very often happens that you have this uncanny feeling of being able to predict another person's behavior, being able to understand another person's actions and you can't explain it. And so you feel like there must be this kind of supernatural force that's connecting you to another person or modulating your behavior, other people's behavior. I don't believe there's any force like that, but I think that's a concept we have to make up to explain these ways in which we engage with other people or with animals or with nature. That we can't understand that on an intellectual level, but are sort of in tune with at a phenomenological level, I guess you can say.

Me: That's great. That's all also interesting to me. I don't have too many other questions. I just was curious if you had anything else to say on the topic, connection between AI, spirituality?

AB: This isn't something I've given a ton of thought to but I've heard some speculation about this. Or just people pointing out that it's that this is going to get weird but it seems like there's a weird space that is going to open up very soon. I mean, it's opening up already and to some extent, I'm sure but for sort of, you know, a AI religion or AI's influence on religion. Already, people are developing personal relationships with chatbots.

Yeah, how long until people start treating them as gods or whatever? I mean, given the way the human mind seems to work, not long. So that's an interesting angle, but you know, obviously, I couldn't even begin to have an idea of what form these sorts of interactions will take.

But a theme, I think that's gone through our discussion is that human beings have this need for a certain type of meaning, a certain way of making sense of things that are intelligible to them. And we fit phenomena into sort of natural schemas and one of those schemas that seems to be very deeply ingrained in human culture and human psychology is the schema of a religion or and or the supernatural. So it seems like a safe bet that AI is going to interact with that in a very deep way.

But kind of in an interesting way, where like, I don't think people are going to be tempted to think, well, I don't know, but it doesn't seem like people are going to be tempted to think of AI as supernatural. That doesn't quite seem like an idea that has a lot of pull. But on the other hand, it does seem like our religious impulse can easily find expression, especially if these systems become more advanced than us, which is something that a lot of people believe will happen in the coming decades.

Then we'll have this weird situation where, in a certain sense, we have created a God. We've created an intelligence greater than ourselves. And so it seems only natural to think that as that process is happening, people are going to start to worship that or to idolize that. So that's all very speculative, but I think it connects up with some of the things we've been talking about.

Me: Yeah, it really does. And I hadn't thought about it from that angle at all. So thank you.

ANAMNESIS

M C M A N U S

Thomas Allen Harris, critically acclaimed, interdisciplinary artist who explores family, identity, and spirituality in a participatory practice. Professor in the Practice at Yale University, Founder of the Family Pictures Institute for Inclusive Storytelling.

Me: You mentioned that you're getting acquainted with AI...

TAH: Yes, I'm getting acquainted with AI. So that means I'm just examining where I encountered where and how I encounter AI without intentionally necessarily using it as a tool myself.

First of all, I should say, last year, I think it's around this time, in the fall of last year, I did a workshop for filmmakers at Google. So three or four filmmakers, diverse filmmakers. And so we did a workshop, exploring how AI with regards to media has advanced.

 And how it's advanced over a 10 year span. And then over the year by year in the last seven, eight years. And so we got a chance to actually see a visual representation of how it's advanced in terms of its construction of fake images and fake people. You could call them fake or AI generated, and so on.

So I decided to invite the Google AI folks to my archive, aesthetics and community storytelling class. And so in my class, I had about maybe 16, 17 people. And three folks came.

Most of the people in my class, many of the people are first gens. So I'm not sure if it's because he is an African American Studies and I'll say film and media studies. So I had a lot of first Americans and people in their first generation of college. Some of them are either or, and then and then there were some people who were multi-generationally American and, multi-generationally had gone to college.

We had two people I think who were born in Africa, Ethiopia, Morocco. I had people from Mongolia, and then a couple of people from Bangladesh and other places, and a bunch of queer and non-binary people. I had a lawyer, an African American biracial lawyer, and then somebody , from architecture and a bunch of other places, in terms of graduate students.

So particularly the people who hailed from other places, like who are part of a larger diaspora, they found themselves frustrated because AI rendition of their home, their home countries, were just like, really kind of like our kind of colonial kind of narrative. These very poor and dirt roads, and, and they were like, we couldn't find a way to construct a place in Bangladesh that was that we knew or where we raised in Ethiopia that we know. It was like starving kids and all that stuff. So that was very interesting to actually have that kind of perspective, because we were generating a kind of collage. With a house and stuff with these prompts.

It's very interesting to be thinking about it, having had that experience in terms of the media imagery. And then you were speaking about writing. So I have used AI with my team in terms of writing letters to funders.

Both: Laughing.

TAH: Letters and also first verse versions of proposals. And so it's interesting to see how certain, like, that kind of communication was, it's, it seemed like, it really shone in that thing. Like, I want to kind of pump you up a little bit. And also pump up my project, you know, that kind of thing. And it hits in certain words, and you get a spotlight kind of thing. And so I did it with other people. It's not my go-to.

But it's been interesting in terms of working on the "My Mom, the Scientist" film, because I'm writing that film now. That's a very personal film, as is a lot of my work, how much I feel like, I don't know, if it's because I've been doing this for 30 years, or also that I've been working on the film for many years now. But I feel like, alright, that I've also been working on the film for many years now. Not exclusively, but with a steady focus.

Right out of the box, I'm very attuned to hitting certain emotional pivots and moments, and combining it with other people's stories, but having a certain anchor. And I'm very clear on how I feel—this film is the first where I ever wrote the entire treatment first.

But in terms of writing my voiceover, and just writing it, it feels like relative to what I've done in the past, I am very clear on what my personal narrative is, and, and how to direct as a driver on the film. And I was just wondering, now that I'm talking to you about this, how much that has to do with AI being able to do a kind of base level? Knowing that that's possible, even though I haven't used it at all.

The truth is, when I see videos on YouTube, now I'm starting to see more AI generated videos that are not advertised as such, but I can tell which ones are AI videos. And I'm like wow, it's so obvious to me in terms of the tone. And it's a lack of various emotional registers.

Me: Yeah, totally.

TAH: So I'm seeing that, and I don't think it can really be replicated, you know? So it's just very interesting to have this kind of experience.

Me: You know, it's so interesting to me how you said it's immediately perceptible. A bunch of other people have said that I completely agree. I feel that a part of my motivation in this project is investigating how, when so much AI generated media is out there, I think we feel it. I think there's some kind of echo effect, or it affects us. So you saying that just makes me think a little more about that.

TAH: Is there a lot out there now?

Yeah. Especially on social media. I haven't seen an AI produced film. I've heard really bad AI produced songs, but yeah, on social media a lot. And, you know, a bunch of people even said to me that they do find it difficult to determine whether something they're looking at is real or fake. This is definitely something that's happening right now. And it has a kind of disorienting effect.

TAH: Yeah, it does, disorienting, and also distancing for me.

Because I've been consuming a lot of news, you know, in the context of the wartime moment we're in, in the Middle East in particular. And I see stuff like "learn how we got here" or "how did this situation arise?" and it's these generic images, you know that that are connected. And then there's this kind of...voice. And then today there was a woman who was talking, and I was like, I was like, I don't know who you are...I don't know what you are!

Both: Laughing.

TAH: And I was like, I'm not going to listen to you anymore!

Me: How can I trust you?

TAH: It's like a different species, right?

Me: Absolutely. I have a friend who's a philosophy professor, actually. And I was having a really interesting conversation with him earlier. He studies theory of mind. And you know, some people even talk about AI by comparing it to an animal intelligence, because it's not a human intelligence, it doesn't think the way a human thinks.

TAH: And then I add to that, like, you know, who's programming it? And what biases does it have, in the way emotions are gendered? And the gendered aspect of emotional intelligence and expression. And then, not to mention, as I indicated earlier, the first world, US bias, and the racialized bias. And it's subconscious in a certain way.

Me: And a big part of my project is investigating the fact that there's actually all this hidden labor in the Global South thousands and thousands of people every day are doing data labeling for AI needs a lot of data, human beings in the global south for really pitiful wages, are the ones looking at language and categorizing language. And this is all sort of, like, cloaked by a machine.

And I think, even beyond the fact that AI can spit out racist stuff, essentializing stuff, very colonial stuff. It's also in the here and now, like, in terms of people's day to day life, a huge, weird, hidden thing— the fact that human beings are actually behind a lot of things that are powering AI, and people don't know. They just perceive it as a machine.

TAH: Interesting. Yeah, it's also thinking about that in relation to what's happening in the Congo, where people who are connected to the Congo or are aware of it, are trying to pierce through,
 the narratives about the dominance of the Israel attack on Gaza, or the war in Ukraine, Russia's war in Ukraine. What's happening in the Congo is fueled by parts that are needed for cell phones.

People are like working, and also being attacked and destabilized. The whole society has been destabilized. And people have been piercing through—not that I do a lot of social media, but when I do, I'm encountering people advocating for not buying a new cell phone. Because you know, you might be paying for it.

And then on top of that, I was listening to a podcast last night on WBAI, it's this guy called Patriotic Millionaire. He was talking about how all these companies like Apple are sitting on so much money and they don't give it out anywhere. They just like parking it, basically. And at the same time, they're not paying taxes in this country. And, and then, creating these foundations that don't give money out. And we're basically underwriting it, and then obviously they're exploiting this Central African country where they get paid pennies to mine, in horrible mining conditions. So when you were talking about the data mining, you know, and and in the relationship between that and the real world— the mirroring of those two is very interesting because we're consuming AI through these phones.

Me: Wow, that's, that's a really fascinating mirror. It's conjuring different images of it now, and I am so appreciative of you, including that in our conversation. Thank you.

MCMANUS

Michael McManus

Me: Who are you?

MM: I'm Michael McManus. I'm a postdoctoral researcher at the UC Berkeley Space Sciences Lab.

Me: True. Ok, so what do you think about AI?

MM: Hmm. Well, just in terms of AI like all the recent LLM, generative AI hype stuff, I would say I'm kind of conflicted. On the one hand it's amazing, mind-blowing technology, like developments have happened so fast that things are possible now that would have seemed completely impossible only 2 years ago. On the other hand, there's so much about it all that is so awful, and there are so many real-world harms happening right now that I feel like are being wilfully swept away and ignored in this tech-hype bubble we're currently in. Although there are lots of people out there fighting the good fight trying to bring attention to these issues and dispel some of the hype!

Me: What do you think is the bad stuff about it?

MM: The fact that it takes legions of underpaid workers in the Global South, often in awful working conditions, to train and fine-tune these models. These are the real ghosts in the machine, these faceless and distant, to us, people whose countless hours of labor are actually what enables these gleaming technological monstrosities to be birthed. The horrible irony is that it seems to me that they're not really being used for that much productive stuff, are they? I can count so many more bad things they're being used for off the top of my head than I can good things.

I've lost count of the number of stories about some newsroom or website firing all their writers to try and lean more into AI, only for them to be called out a few weeks later for producing error-ridden AI generated slop that no one wants to read, most recently the Sports Illustrated thing. They're being used for creating better looking spam and misinformation on completely unprecedented scales. They seem to be amazing tools at actively making the internet a worse place - like all those recent examples of AI generated images polluting Google's image results? Or just straight up misinformation at the top of the search results? Or how it's enabled deepfake AI porn like never before. I think there was a Bloomberg piece about that a few days ago. Feels like these things are only going to get much worse. And like what is the point of all of this stuff? How is this making anyone's lives better? Is all this really worth it, just so that we can write a boring email slightly more quickly?

And there are ever more papers coming out now about the environmental impact of these LLMs and generative AI models. Both in terms of the obviously huge amounts of electricity needed during the training phase, but also the power and water needed on an individual query level. It's kind of shocking! I feel like I'm getting horrible flashbacks to the crypto hype bubble.

Me: What, you don't love crypto?

Both: Laughing.

MM: Another thing that's stood out to me is just how rushed and crap so many of the AI products so far have been, like Microsoft's Bing AI and Humane's AiPin both having glaringly obvious hallucinations in their launch videos, or Google's AI email assistant just fabricating emails that don't exist, or most recently Amazon's Q AI assistant thing instantly being shown to both leak real confidential data and make things up, which is kind of a hilarious worst of both worlds. Is capital's race to the bottom really so frenzied now as to be rushing out such shoddy product demos? Something about that feels kind of weird and unprecedented to me.

Me: Right. And you mentioned before that to you it's not all bad—what are some of the things that you find more interesting, or like about it then?

MM: When it comes to LLMs, I guess I'm quite interested in their failure modes, things they're not good at. Mainly because they help demarcate the limits of their behavior and help lay bare what is actually going on when you type in a query. I think having as good a mental and intuitive model as is possible for how these things behave is important for seeing through hype and preventing people from anthropomorphizing them, which can be difficult since the quote unquote intelligence they display is so idiosyncratic, and we're really not used to interacting with software that behaves like this.

A good example of this is to ask it a classic

logic puzzle or brain teaser, but change one or two of the details so as to make it trivial to us. For instance, ask it how to measure out 4 liters of water using a 3 liter and a 4 liter bucket. The well-known puzzle gives you 3 and 5 liter buckets. Often it instantly spits out a response that is as if it's been asked the usual puzzle, just with the numbers you gave it. This is because it's seen that puzzle thousands if not tens of thousands of times in its training data, but this modified but trivial version no one's ever thought to ask online, because why would they? And the LLM cannot reason out the fact that the solution is now straightforward.

Another way of thinking about it is that a small change in input, the number 5 going to the number 4, in the real world actually produces a large change in output. I e, the solution is just one step now, but the lossy compression of the world that is an LLM has papered over and smoothed out this kink, with the result being a poor approximation of the world. I don't know, I find very simple examples like that kind of interesting to think about, mainly because they help remind me how these language models work. And that GPT doesn't actually know what a bucket is, even though it might appear that it does 90% of the time.

Another very interesting recent example was a paper showing that if an LLM learns a statement of the form A is B it cannot necessarily deduce the reverse, B is A. For example if you ask GPT who is Tom Cruise's mother, it will say Mary Lee Pfeiffer, but if you ask it who is Mary Lee Pfeiffer's son, it won't know. This reflects the fact that sentences of the form "Tom Cruise's mother is..." presumably appear far more often in the training data than sentences of the form "Mary Lee Pfeiffer's son is..."

One funny thing that you notice whenever things like this go semi-viral on Twitter or whatever is that quite often they seem to get quote unquote fixed pretty quickly, and the behavior goes away. Not to go full on tinfoil hat mode but I really feel like OpenAI hot-patches fine tunings...which I guess is just PR and good for business, but it's definitely an inherently losing and pointless battle. You're not going to whack-a-mole your way to a general reasoning machine like that.

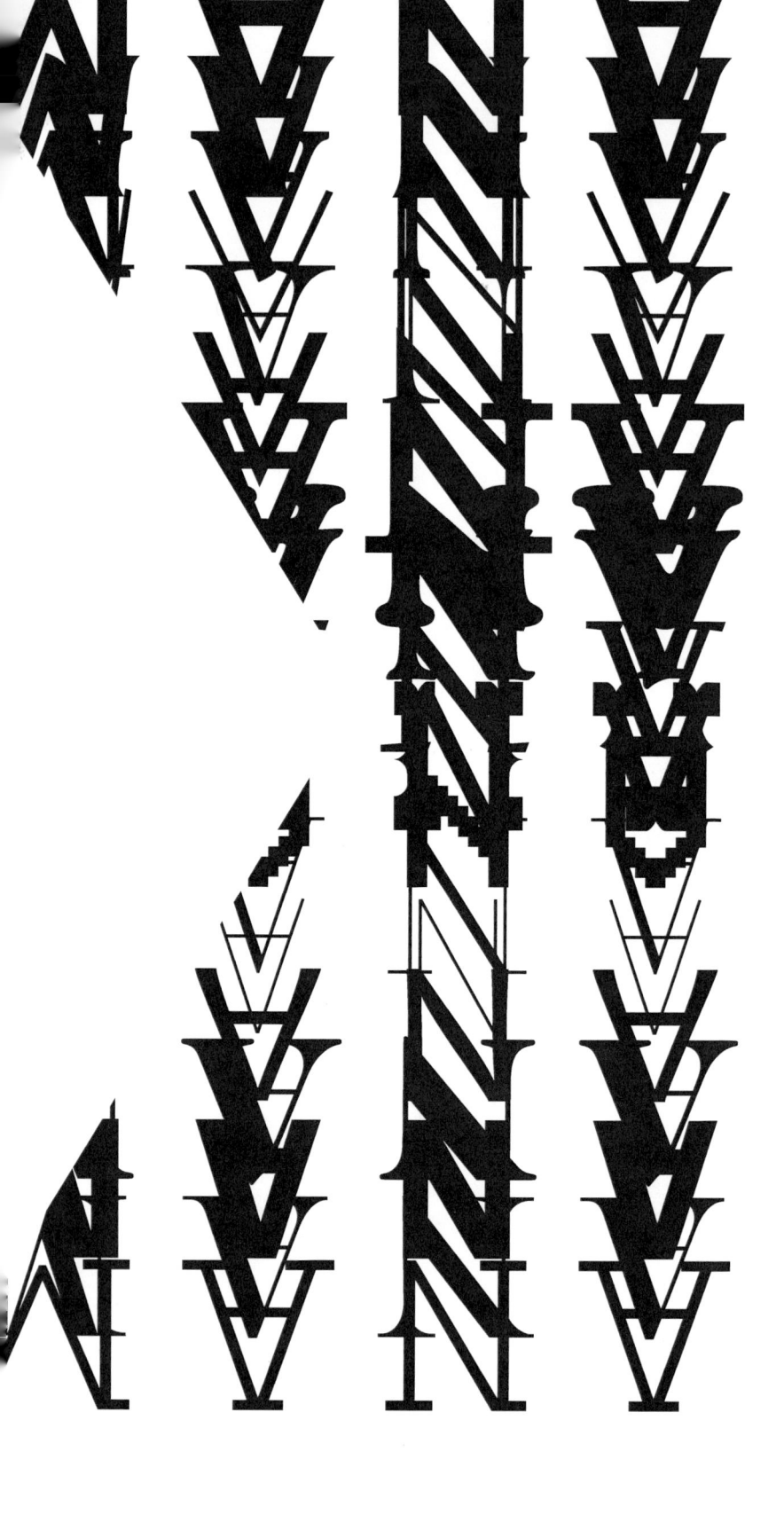

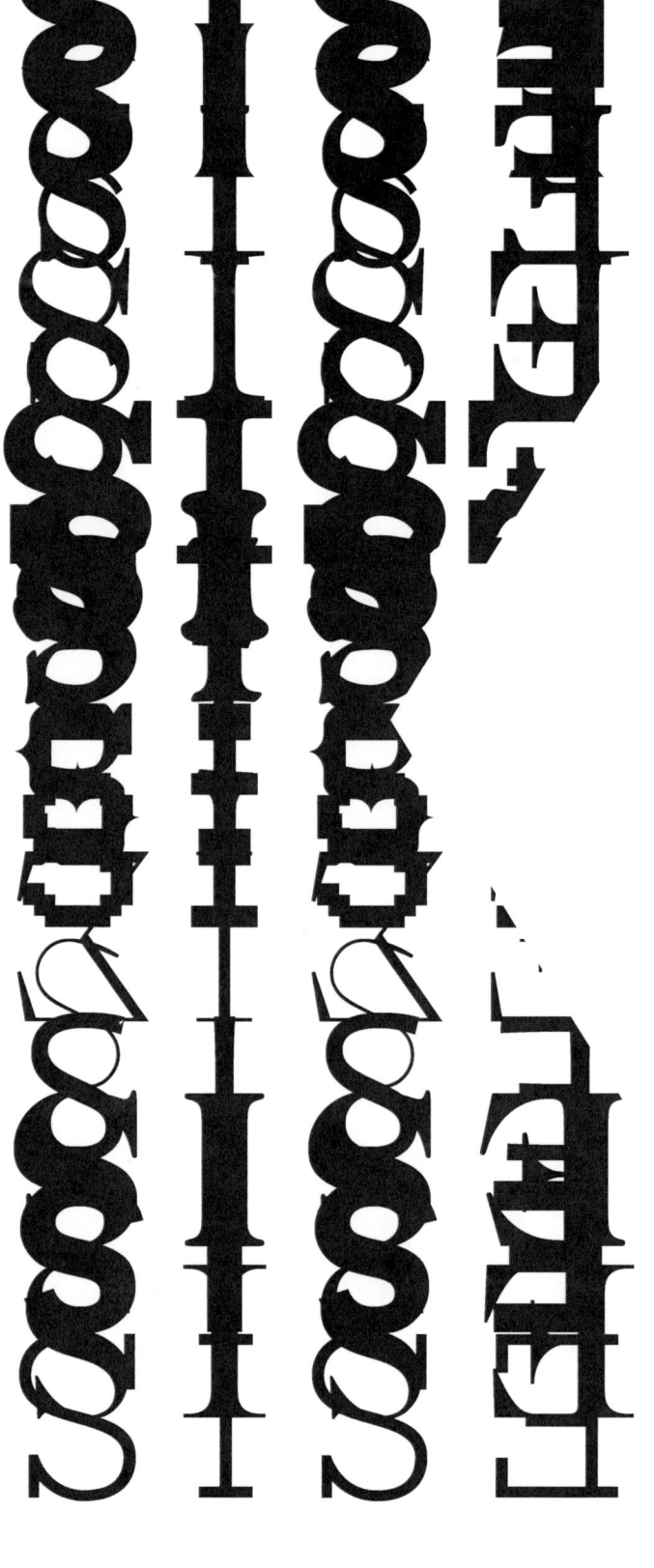

www.ingramcontent.com/pod-product-compliance
Lightning Source LLC
Chambersburg PA
CBRC091630140626
46547CB00027B/691